Cambridge Imperial and Post-Colonial Studies

The Cambridge Imperial and Post-Colonial Studies series is a well-established collection of over 100 volumes focussing on empires in world history and on the societies and cultures that emerged from, and challenged, colonial rule. The collection includes transnational, comparative and connective studies, as well as works addressing the ways in which particular regions or nations interact with global forces. In its formative years, the series focused on the British Empire and Commonwealth, but there is now no imperial system, period of human history or part of the world that lies outside of its compass. While we particularly welcome the first monographs of young researchers, we also seek major studies by more senior scholars, and welcome collections of essays with a strong thematic focus that help to set new research agendas. As well as history, the series includes work on politics, economics, culture, archaeology, literature, science, art, medicine, and war. Our aim is to collect the most exciting new scholarship on world history and to make this available to a broad scholarly readership in a timely manner.

More information about this series at
http://www.palgrave.com/gp/series/13937

Suryakanthie Chetty

‘Africa Forms the Key’

Alex Du Toit and the History of Continental Drift

palgrave
macmillan

Suryakanthie Chetty
History Department
Stellenbosch University
Stellenbosch, South Africa

ISSN 2635-1633 ISSN 2635-1641 (electronic)
Cambridge Imperial and Post-Colonial Studies
ISBN 978-3-030-52713-6 ISBN 978-3-030-52711-2 (eBook)
https://doi.org/10.1007/978-3-030-52711-2

Cover illustration: Portrait of Alex du Toit, courtesy of University of Cape Town Libraries (Jagger Library)

This Palgrave Macmillan imprint is published by the registered company Springer Nature Switzerland AG.
The registered company address is: Gewerbestrasse 11, 6330 Cham, Switzerland

Preface and Acknowledgements

When I was very young, books and babysitters were in short supply. My mother, being rather innovative as is the wont of single, working mothers, saw the municipal library as a solution. Every Saturday morning, armed with a sandwich and a few cents, I would find my favourite spot on a corner bench by the window of the children's library and pore through the books. While television had reached the shores of South Africa a few years previously, it had yet to reach our home so reading was the entertainment of choice. In a happy coincidence then, the interests of my parent and myself intersected.

My reading was incredibly varied, ranging from the science fiction of Robert Heinlein to classic adventures of the nineteenth century, be it *The Lost World* of Arthur Conan Doyle or Jules Verne's *Journey to the Centre of the Earth*. And, in a desire to learn more about the history of the world—even then, a passion—I found myself drawn to the non-fiction section. The books that inevitably caught my attention were those depicting seismic activity in glorious technicolour. Living in the relatively seismically stable Durban, the world of volcanoes, earthquakes and horrifyingly fascinating tsunamis challenged my preconceptions of the ground on which I walked. With the carefree disregard for copyright that characterises an eight-year-old, I would use my limited funds to photocopy five pages a week of a book that had particularly caught my eye. Photocopied in grainy black and white on A3 paper, the sheets would then be rolled up and secured with a rubber band to be perused at leisure when I was at home. A few years later when we had an all too brief encounter with geology in a geography class; I needed only to hear the date "1873" to know that it represented the cataclysmic eruption of Krakatoa in Indonesia or that 1906 was the year of the San Francisco earthquake. It is my unfortunate confession as a historian that those still tend to be the few dates that I remember with any clarity.

I must have been about eight years old the first time I recognised a certain wonder that comes with understanding a part of the natural word and our place in it. This moment of reverential clarity came to me while watching a re-run of Carl Sagan's television series *Cosmos* and hearing him enunciate in his distinctive tones that we were "star stuff"—composed of the atoms forged at the fiery heart of distant stars and released into the universe at the moment of their death.[1]

Scientific thought is characterised by many such moments of poetry—Newton's theory of gravitation that explained everything from falling apples to orbiting planets, Einstein's theory of relativity that addressed the counterintuitive subatomic world, Darwin's theory of evolution that accounted for the beauty and complexity of life. For me, plate tectonics in its elegant simplicity evoked the same awe as the realisation that I was made of "star stuff".

It was based on my interest in and fascination with plate tectonics that I first heard of Alex Du Toit, a South African geologist, who was incredibly farsighted in his early advocacy of a theory that would only gain mainstream recognition after his death. He became the ideal conduit for telling the story of plate tectonics—and geology—in South Africa.

This work then, despite the numerous detours and digressions that characterise one's life, seems to be a destiny made manifest. It is a product of a lifelong infatuation with both history and geology, our interaction with each other and the world around us, the factors that have shaped our past and influence our future.

This work has been the product of the support, encouragement and assistance of a number of people:

Craig Smith and the Geological Society of South Africa (GSSA) Archives, Estelle Grobler at the Council for Geoscience Library and Information Centre and, in particular, Clive Kirkwood and his staff at the Jagger Library, University of Cape Town.

[1] Carl Sagan, Ann Druyan and Steven Soter, *Cosmos*, Gregory Andorfer and Rob McCain, producers, 1980.

Keith Breckenridge, Catherine Burns, Marijke Du Toit, Jeff Guy, Julie Parle and Goolam Vahed who were instrumental in fostering my love for history; the History Department at the University of Johannesburg and, in particular, Gerald Groenewald and Karie Morgan for their unwavering support; Melanda Blom and my mentors, Tilman Dedering and Alex Mouton at the History Department at Unisa; my new colleagues at the History Department, Stellenbosch University; the geologists who tolerated my enthusiasm with great forbearance: Bruce Cairncross and Elna van Niekerk; and to John Connolly whose words inspire me.

Then there have been those who have been my pillars of strength and my champions: Prinisha Badassy, Anne and John Samson and Nicholas Southey.

Finally, I wish to dedicate this work to my mother with my love and gratitude…always.

Stellenbosch, South Africa Suryakanthie Chetty

Contents

List of Figures

CHAPTER 1

Introduction

Geology—while being firmly based in the sciences—can be considered an historical science. It reveals a history of the planet that, unlike conventional history, is not centred on humankind. In the geological view of "deep time", the emergence of humanity accounts for only a minute portion of the earth's history. Yet, simultaneously—and perhaps contradictorily—geology and history cannot be divorced from each other. The fairly new field of "Big History" draws upon both the sciences and the social sciences in order to create a new and more encompassing view of human history. This is evident in *Origin Story* by David Christian, *Big History* by Cynthia Stokes Brown and, most recently, *A Most Improbable Journey* by geologist Walter Alvarez which offers an engaging introduction into the field from a geological perspective.[1] Despite a tendency to marginalise human agency, the notion of "Big History" offers an understanding of the ways in which larger environmental factors have shaped the course of human history. Geology has an important part to play here—the ground beneath our feet has an impact upon the type of homes we build, the food we grow and the trade in which we engage. This has in turn played

[1] David Christian, *Origin Story: A Big History of Everything* (Allen Lane, 2018); Cynthia Stokes Brown, *Big History: From the Big Bang to the Present* (New York, The New Press, 2012); Walter Alvarez, *A Most Improbable Journey: A Big History of Our Planet and Ourselves* (New York, W. W. Norton and Company, 2017).

S. Chetty, *'Africa Forms the Key'*, Cambridge Imperial and Post-Colonial Studies, https://doi.org/10.1007/978-3-030-52711-2_1

a role in the rise and fall of civilisations, wars and imperialism, the emergence of equality and inequality.

Yet geology, as a discipline and a means of understanding the processes of the earth, also has a history—a history that has been shaped by the practitioners of the discipline and the ideological worlds that they inhabited. This history of the discipline is one that has been shaped by controversy and consensus, demonstrating that the path to "truth" does not run smoothly. This work concerns itself with one such controversy—continental drift. In the mid-twentieth century, the discipline of geology underwent a seismic shift with the recognition of plate tectonics. This is based on an understanding that the earth's crust (both continental and oceanic) is divided into a series of segments or plates that, powered by convection from the earth's interior, move along the surface of the earth. These movements bring plates into contact with each other leading to the formation of mountains and the destructive power of earthquakes and volcanoes. Plate tectonics has therefore become fundamental in understanding the earth's geology both in the present and in the past and is a means of attempting to predict the future.

The origin of plate tectonics lies in the theory of continental drift that, despite having antecedents, was particularly prominent as a subject of dissent in the decades leading up to the Second World War. The view that the ground on which we stand marked by the boundaries that serve as markers of our cultural and national identity and a symbol of permanence was in constant motion challenged the ways in which geologists saw the world and their place in it. Reactions were extreme ranging from those who enthusiastically accepted and validated this view with the evidence that they found all around them to those who vehemently opposed it and summoned different interpretations of the same evidence.

Alex Du Toit was one of the figures at the forefront of this controversy. Arguably one of South Africa's greatest geologists with a formidable understanding of the country's geology, Du Toit is most remembered in international circles for his work on continental drift. As a drift theorist, scientist and intellectual, he also serves as a representative of a significant time in South African history defined by both incorporation and exclusion. South Africa was a newly independent nation that had been shaped by imperialism and had, in turn, embarked on policies of segregation and discrimination designed to limit the idea of the nation. Du Toit's work on shifting spaces therefore exemplifies the contradictions and complexities of pre-apartheid South Africa. This project thus uses the figure of Alex Du

Toit to explore the ideological milieu in which scientific knowledge is constructed, both internationally and domestically. It demonstrates that, while conventional histories of continental drift acknowledge Du Toit's work in Africa to an extent, Du Toit's various studies of the regions that would comprise Gondwana—with an emphasis on Africa and South Africa, in particular—and the conclusions he drew from this were key in providing the evidence for continental drift. Simultaneously, Du Toit's work was contextualised by South Africa's growing assertion of African centrality in scientific discovery, evident in the fields of palaeoanthropology and geology that challenged the intellectual dominance of the northern hemisphere, even as it drew upon this tradition.

The history of geology falls within the larger field of the history of science. In the South African historiography the history of science is a relatively new field and, understandably, given the country's turbulent past, much of the work done on the history of science in South Africa focuses on its relation to ideas of race. In *Scientific Racism in Modern South Africa*, Saul Dubow highlights several themes which are of benefit to this study. Key here is the ideological and political context which frames the acquisition of scientific knowledge where fields such as physical anthropology and archaeology accounted for human origins and settlement patterns in a manner imbued with contemporary concerns regarding racial distinction and hierarchy.[2] Recently, *Darwin's Hunch* by Christa Kuljian seeks to focus on the interaction between racial ideology and scientific thinking in the palaeoanthropological discoveries that have taken place in South Africa over the past century.[3]

Further, Saul Dubow's *A Commonwealth of Knowledge* addresses the intricate and often fraught relationship between South Africa and Britain. This is a work that focuses on the white intellectual elite as they used Enlightenment values to create institutions such as museums and universities as well as a public intellectual culture as a means of affirming South African national identity and claims for political rights in the face of British imperial domination. While Britain—and Europe—remained the source of intellectual expertise, the transmission of knowledge was not a unilateral one but based instead on the movement of people and information

[2] Saul Dubow, *Scientific Racism in Modern South Africa* (Cambridge, Cambridge University Press, 1995).

[3] Christa Kuljian, *Darwin's Hunch: Science, Race and the Search for Human Origins* (Johannesburg, Jacana Media, 2016).

within the Commonwealth and adapted to local needs and aspirations. In South Africa, following independence, science—with its associations of neutrality and progress—underpinned political rule and racial exclusion. It would later be harnessed to the project of Afrikaner nationalism by the apartheid state as the country was increasingly isolated. With the advent of full democracy in 1994, Thabo Mbeki's "African Renaissance" became another means of demonstrating both a national and Africanist identity that privileged indigenous knowledge systems. By tracing the shifting uses of knowledge through South Africa's history, Dubow is able to situate scientific thought within a local, national and international framework that challenges one-dimensional notions of scientific history.[4] The story of Alex Du Toit and continental drift is therefore situated within the existing literature, with a particular focus on the ways in which geological knowledge was constructed in South Africa.

An important figure within the history of geology in general is that of Martin Rudwick. An early work of his, *The Great Devonian Controversy* published in 1985, discusses the heated debates within the Geological Society in Britain in the first half of the nineteenth century over the stratigraphic classification and relative dating of floral fossils found in Devon, England.[5] It is significant in demonstrating the ways in which the construction of scientific knowledge was not a neutral process but based on the preconceptions of scientists which were not always easy to abandon. The heated arguments between geologists here evoke that between continental drift proponents and opponents a century later. Rudwick has achieved further renown with two further volumes which provide a comprehensive account of the history of modern geology. Extensively researched and offering a wealth of detail, *Bursting the Limits of Time: The Reconstruction of Geohistory in the Age of Revolution* discusses the very beginnings of geology with its challenge to biblical notions of time and the centrality of humankind. Rudwick traces the piecing together of knowledge derived from the work of amateur and professional geologists from the eighteenth century in presenting a radically different view of the history of the earth.[6] *Worlds before Adam: The Reconstruction of Geohistory*

[4] Saul Dubow, *A Commonwealth of Knowledge: Science, Sensibility and White South Africa, 1820–2000.* (Oxford, Oxford University Press, 2007).

[5] Martin J.S. Rudwick, *The Great Devonian Controversy: The Shaping of Scientific Knowledge among Gentlemanly Specialists* (Chicago, University of Chicago Press, 1988).

[6] Martin J.S. Rudwick, *Bursting the Limits of Time: The Reconstruction of Geohistory in the Age of Revolution* (Chicago, University of Chicago Press, 2007).

in the Age of Reform continues the story of the history of geology looking at dissent and debates amongst geologists as the eighteenth gave way to the nineteenth century. Again, these debates were contextualised within the existing religious and ideological beliefs of these figures and evident in, for instance, the conflict between catastrophists and uniformitarians over the geological processes that shaped the earth.[7]

Rudwick would also assess the work of historian Roy Porter whose work marked a major departure from contemporary histories of geology that were usually the preserve of geologists and tended to emphasise individual figures and the ways in which their work was influenced by the relationship between theory and observation. Porter moved beyond what geologists were saying to the ways in which this was articulated and what this revealed about them and the societies that produced them. Porter demonstrated that James Hutton had employed a deductive approach, only engaging in fieldwork after the formulation of his theory and in order to support it. In contrast, Charles Lyell—while following in Hutton's footsteps—would adopt a more inductive approach. The history of geology became the means by which Porter was able to argue that science was not "natural" but "constructed".[8] The ultimate triumph of uniformitarians in the persons of Hutton and Lyell would have an impact on the way in which continental drift theory was initially perceived.

The most comprehensive account of the history of continental drift theory is Henry Frankel's four-volume work, *The Continental Drift Controversy*. Based on an extensive use of primary oral and written material, Frankel's account is rich and detailed as it traces the history of drift theory over the course of the twentieth century across much of the world. The first volume begins with Alfred Wegener's theory and the subsequent controversy that had the effect of splitting geologists between drift proponents and opponents. It forms the context for my own work on Alex Du Toit who was both a contemporary of Wegener and a figure thrust into the midst of the controversy. Subsequent volumes address the validation of drift theory in the mid-twentieth century and the transformation of drift

[7] Martin J.S. Rudwick, *Worlds before Adam: The Reconstruction of Geohistory in the Age of Reform* (Chicago, University of Chicago Press, 2010).

[8] Martin Rudwick, *Roy Porter, Historian of Geology* in *History of Science*, xli, 1 September 2003; Roy Porter, *The Making of Geology: Earth Science in Britain, 1660–1815* (Cambridge, Cambridge University Press, 1977).

theory in plate tectonics—a view of the earth's geological processes that underpins geological thinking today.[9]

Plate Tectonics: An Insider's History of the Modern Theory of the Earth edited by Naomi Oreskes addresses various aspects of the history of plate tectonics from its origins in continental drift theory to the ways in which developments were contested by geologists. It is an important text in showing the ways in which geological—and scientific—knowledge is contexualised by time and space. Of note is an essay by geologist, Frederick Vine, one of the pioneers in plate tectonics theory in the 1960s.[10] Naomi Oreskes' *The Rejection of Continental Drift* also provides important insight into the way in which scientific methodologies—which are employed and prioritised in different ways—affect the ways in which knowledge is created and received. This held true for continental drift where the antagonism of American geologists was based on a different use of the scientific method that was in contrast to that employed by Alfred Wegener.[11]

The primary material on which this work is based draws largely upon the Alex Du Toit papers housed at the University of Cape Town Libraries. Du Toit was a prolific note-taker, using a small and often all but illegible script, accompanied by pencil sketches of geological formations. He also kept records of correspondence as well as newspaper articles that pertained to his work or caught his interest. At the same time, there is a dearth of personal information which is reflected in this work. As such, this is less a biography of Alex Du Toit than it is an account of the history of continental drift and his particular positioning as a South African geologist. It is through his work and his enthusiasm for an unpopular and controversial theory that some inkling of his character is evident.

While this work is structured as a broad chronology, the relevance of particular themes and the need to provide both context and background has sometimes necessitated a departure from the strict chronology. In addition, chapters focusing on Alex Du Toit are interspersed with chapters that detail the history of continental drift and geology in general for the lay reader. Furthermore, while there is a brief consideration of the remainder of Du Toit's prolific geological career, a detailed discussion is beyond

[9] Henry R. Frankel, *The Continental Drift Controversy*, Vols 1–4 (Cambridge, Cambridge University Press).

[10] *Plate Tectonics: An Insider's History of the Modern Theory of the Earth*, Naomi Oreskes, (ed), (Boulder, Westview Press, 2003).

[11] Naomi Oreskes, *The Rejection of Continental Drift: Theory and Method in American Earth Science* (New York, Oxford University Press, 1999).

the scope of this project. The focus on Du Toit's life is therefore largely limited to continental drift theory and its impact.

Chapter 2 addresses the growth of scientific knowledge and geological exploration in colonial South Africa in the nineteenth century. In southern Africa, as in much of the rest of the world, geology was harnessed to the imperial project in terms of the geological mapping and search for mineral resources in the far-flung territories of the globe. The late nineteenth century in South Africa also saw the growing professionalisation of geology that privileged the learned geologist rather than the adventurous amateur. Accompanying this was the establishment of professional organisations and the discipline itself gained both significance and impetus through the discovery of mineral resources that would come to shape the country's history. It is this that contextualised the birth and early education of Alex Du Toit and the subsequent focus here is on his education in Britain and his significant work in geological mapping as a member of the Geological Commission of the Cape of Good Hope—work that would form the basis for the controversial stance that he would take subsequently.

Du Toit's education in Britain—and in Scotland in particular—highlights the relevance of this part of the world in relation to the growth of modern geology as a discipline. This is the focus of the third chapter which looks at the significance of James Hutton and, subsequently, Charles Lyell in laying the foundations for modern geology. Their emphasis on uniformitarianism—or gradual and steady geological change—challenged catastrophism with its religious associations and emphasised the great antiquity of the earth. The increasing disavowal of catastrophism, however, set the scene for the conflict that would arise as a result of continental drift with its erroneous associations with dramatic change. While associated with the early twentieth century, this chapter also briefly addresses early notions of continental drift occurring within the British imperial context with a particular focus on South Asia.

Chapter 4 returns to South Africa with attention paid to the relationship between British imperialism, South African nationalism and geology. Beginning with the first visit of the British Association for the Advancement of Science in 1905, the chapter relates the complex history of geology and mining in southern Africa. This would influence the independent nation that came into being in 1910. Du Toit's early work in geology illustrated the two worlds that he straddled—that of empire and the nation. With a firm grounding in the scientific tradition that arose out of the Enlightenment, Du Toit's work was nevertheless focused on South Africa and a growing understanding of its importance in continental drift. His work in the early years after the formation of the Union of South Africa and while serving during the First World War, allowed for the embryonic development of continental drift theory.

Chapter 5 deals with the history of continental drift theory that contextualised Du Toit's work—even as he contributed to it. It addresses the early work of Eduard Suess in the nineteenth century and, more significantly that of Du Toit's contemporaries, Frank Bursley Taylor and Alfred Wegener in the early twentieth century. The latter, in particular, aroused the ire of geologists committed to the notion of the immobility of the earth's surface and it is this extreme feeling on the part of both "mobilists" and "fixists" that demonstrate the ideological construction of scientific knowledge. This chapter also briefly considers the exchange of information and interaction between Taylor, Wegener and Du Toit which is suggestive of the latter's importance as a key figure in the debate.

The following chapter focuses on the political climate of segregation and liberalism in 1920s South Africa. This is evident at a meeting of the South African Association for the Advancement of Science held in Durban in 1921. It was here that Du Toit delivered a public lecture that explicitly dealt with continental drift, drawing together evidence to argue for the prior union of the southern landmasses in the form of the supercontinent, Gondwana. This was followed by the publication of "The Carboniferous Glaciation of South Africa" that thrust Du Toit into the international arena as an important drift theorist. This chapter draws attention to the ways in which Du Toit was articulating a new—and controversial—concept of space just as South Africa was embroiled in internal struggle over space, epitomised by segregation.

Du Toit's reputation would be established with the publication of *A Geological Comparison of South Africa with South America* published in 1927. His work was based on the research undertaken during a trip to South America in 1923, made at the behest of and with the financial assistance of the Carnegie Institution. This forms the core of Chap. 7 which provides a summary of Du Toit's published research. Du Toit's use of geological stratigraphic and fossil evidence would be a pattern set for future work in continental drift. Through these means he was able to reconstruct the ancient continent of Gondwana comprising South Africa, South America, India, Australia and Antarctica. Du Toit's brief period of research in South America also meant a reliance on the work of like-minded South American geologists, leading to the formation of relationships that were marked by both co-operation and dissent. This chapter also considers Du Toit's role as a teacher and work in the mining industry, aspects of his multifaceted career that were nevertheless shaped by his interest in continental drift.

Du Toit's research in South America and conclusions regarding continental drift occurred within a significant period of scientific development and discovery in South Africa that is addressed in the subsequent chapter. The year 1924 was a key one. Jan Smuts had lost the election and, with an enforced period out of politics, worked on his scientific philosophy of holism. At the same time, Raymond Dart made a discovery that would challenge existing understandings of human origin. The Taung skull was that of an *Australopithecus africanus*, an early hominin, and suggested the country's importance as a site of human evolution. This centrality was affirmed by Du Toit's own work which implied the country's geological significance in the reconstruction of Gondwana. With the support of Smuts and J.H. Hofmeyr, the intellectual Administrator for the Transvaal, these discoveries were an assertion of nascent nationhood that allowed a greater confidence when the second visit of the British Association for the Advancement of Science marked the close of an eventful decade.

In 1937 Du Toit would publish *Our Wandering Continents*, a landmark work that marked the culmination of his thinking on continental drift. It would also generate substantial hostility on the part of geologists—drawn largely from the northern hemisphere and the United States of America in particular—that would highlight the ideological, divisive and controversial nature of the continental drift debate. Described as a "dense" but nonetheless formidably argued and detailed thesis, *Our Wandering Continents* is briefly summarised in Chap. 9 with, however, greater attention paid to the responses it evoked. At the heart of criticisms of continental drift theory was the lack of an explanatory mechanism for drift. This, too, was a weakness of Du Toit's work although, like Wegener and Taylor before him, he did attempt to suggest the possible mechanisms for motion. While valid, criticisms of continental drift theory—as well as the responses by drift proponents—were shaped by an ideological milieu that was inextricably linked to space. This occurred in the form of assertions of national identity but was also particularly related to the flow of knowledge along the north-south divide.

Chapter 10 focuses on the Antarctic—a geographical area that was highly significant for the evolution of continental drift theory. Floral and faunal evidence was accumulated by early expeditions to the hostile region and the finding of *Glossopteris* fossils by the tragic Scott expedition as well as other fossilised matter indicated that the Antarctic had had a more temperate climate in the past. Various mechanisms were used to explain this change in climate including that of "polar wandering" and, of course, the movement of continental landmasses. The chapter then discusses Du Toit's attempt to mount a South African research expedition to the Antarctic, an important site for drift theorists, rendered even more so by its inaccessibility.

Chapter 11 focuses on the final years of Du Toit's life and legacy, even as drift theory remained controversial.

"Conclusions" contextualises the history of continental drift and Du Toit's work in particular within changing and contested notions of space. Scientific endeavour—with its aspirations to truth—is not value-free and this is highlighted by continental drift which spoke to the power relations evident in the transmission of knowledge between the northern and southern hemispheres, the context of colonialism, war and segregation.

CHAPTER 2

In the Beginning…: Geology in South Africa and the Early Years of Alex Du Toit

At the time of Alex Du Toit's birth in the late nineteenth century, the territory that would become South Africa in 1910 had undergone a period of important intellectual development contextualised by its position as both a colony and emerging nation. This was compounded by the mineral discoveries of Kimberley and the Witwatersrand and their implications for the study of geology. And Du Toit's history was emblematic of this new nation. Of French Huguenot descent, his ancestry dated to Francois Du Toit, originally from Northern France, who arrived in South Africa in 1687. The Huguenots were Protestants fleeing religious discrimination in France and the Dutch East India Company (VOC or Vereenigde Oostindische Compagnie) settlement at the Cape provided a haven. They were incorporated into the fledging Dutch community and given land on which to farm.[1] They would later become identified with the Afrikaner population. Du Toit's middle name, "Logie", however, suggested a more complicated family history. Captain Alexander Logie was a Scottish officer in the Royal Navy who married a Du Toit and subsequently adopted her nephew who was christened Alexander Logie Du Toit. This younger Du Toit married Alexander Logie's niece, Anna Logie. The match produced four children,

[1] Robert Ross, *A Concise History of South Africa, Second Edition* (Cape Town, Cambridge University Press, 1999, 2008), p. 24.

S. Chetty, *'Africa Forms the Key'*, Cambridge Imperial and Post-Colonial Studies, https://doi.org/10.1007/978-3-030-52711-2_2

the oldest of whom was Alexander Logie Du Toit.[2] From the outset, then, Du Toit represented the complex mix of Briton and Afrikaner that would play out in intricate ways in the early twentieth century.

The Evolution of Scientific Knowledge

The 1820s saw a flourishing of intellectual activity in Cape Town with the formation of the South African Library, the South African Institution, the South African College (later to become the University of Cape Town), the South African Literary Society and the South African Museum. The latter was where early geological specimens were housed. From the outset these institutions reflected the exclusions in South Africa with memberships that were largely confined to English and Dutch-speaking middle class white men. This was particularly evident in the case of the South African College. These early institutions can also be contextualised within existing imperial knowledge networks. Andrew Smith founded the South African Museum. Originally from Britain, he mixed both politics and science with his trips into the interior ostensibly for specimens but also to establish lines of communication with African chiefs on behalf of Britain. He was subsequently instrumental in the formation of the South African Institution with its mandate "to stimulate investigation of the geography, natural history, and general resources of the country".[3] Clearly the theoretical aspects of science would not be divorced from the practical applications—which would be particularly pertinent in the case of geology.

The late nineteenth century marked the period of the formalisation of scientific knowledge in South Africa. The South African Philosophical Society was established in 1877. Modelled on similar British Victorian institutions, it was the realm of the amateur scientist, an elite institution made even more so by the £2 membership fee.[4] Straddling the divide between the amateur and the professional was the field of geology. With its origins as a modern discipline in the Enlightenment, geology was

[2] T.W. Gevers, "The Life and Work of Dr. Alex. L. Du Toit"—Alex L. Du Toit Memorial Lectures No. 1. In *The Geological Society of South Africa: Annexure to Volume LII* (Johannesburg, The Geological Society of South Africa, The South African Association for the Advancement of Science, The South African Geological Society, 1949), p. 1.

[3] Saul Dubow, *A Commonwealth of Knowledge: Science, Sensibility and White South Africa, 1820–2000* (Oxford, Oxford University Press, 2007), pp. 36–39.

[4] Saul Dubow, *Scientific Racism in Modern South Africa* (Cambridge, Cambridge University Press, 1995), p. 11.

inextricably linked to the modern world and harnessed to the imperial project. Yet its true origins lay at the beginning of human civilisation.

Early Geology in South Africa

The practical knowledge of geology, ranging from building material to agriculture, the extraction of precious metals and gems to comprehending the nature of earthquakes and volcanoes, means that the science has a long history in various incarnations. An integral part of mythology, various philosophers from the seismologically unstable ancient Greece to Asia have attempted to decipher the way in which the earth was constructed. It was, however, the Scientific Revolution, closely followed by the Enlightenment that gave geology its recognisable characteristics. These did not go unchallenged. The implications of geology for religious belief in terms of determining the age of the earth meant that, just as astronomy had relegated man from his central position in the universe, the emerging science of geology undermined the very ground on which he stood. By the mid-eighteenth century both Germany and France had established mining schools. The Industrial Revolution and its accompanying quest for raw materials in the form of iron and coal provided the impetus. The association of geology with the Industrial Revolution continued in the form of James Hutton, considered to be the father of the modern discipline. His close association with James Watt and knowledge of the steam engine—of which Watt was the architect—informed his description of the seismic processes of the earth.[5] These trends in the acquisition and interpretation of scientific knowledge occurred just as the European world was expanding dramatically. Voyages of exploration and discovery were soon succeeded by early attempts at colonialism.

From the outset of colonialism, the history of geology has been intertwined with the search for minerals. Within three decades of the Dutch arrival at the Cape of Good Hope in the mid-seventeenth century, Governor Simon Van Der Stel authorised the first foray into Namaqualand in search of copper. The challenges of navigating a largely unknown wilderness put paid to prospective mining efforts in the area but this was to set the tone for what was to come. The growing exploration and

[5] Gabriel Gohau, *A History of Geology*, Albert V. Carozzi and Marguerite Carozzi (translators) (New Brunswick and London, Rutgers University Press, 1990), pp. 100, 111–112.

expanding colonialism that characterised South Africa in the early nineteenth century also contributed to a furtherance of geological knowledge of the landscape with descriptive accounts and the collection of samples by travellers and explorers. Geological descriptions became part of the knowledge of empire—the letters of Captain Basil Hall detailing the geology of the iconic Table Mountain was read before the Royal Society of Edinburgh in 1813. Hall was the son of James Hall, eminent geologist who had done pioneering work in supporting the work of James Hutton. In 1836, Charles Darwin stopped at Simons Bay while voyaging on the H.M.S. *Beagle* and was taken on a brief geological tour of the region by Andrew Smith.[6]

These early writings, unsurprisingly, contributed to the development of early geologists, pre-eminent of which was Andrew Geddes Bain.[7] Bain was part of the contingent of British settlers who arrived in South Africa in 1820 and was soon participating in explorations of the interior.[8] After a series of adventures and a rise and fall in fortune, Bain was tasked with the construction of roads under the auspices of the Royal Engineers Department. Inspired by the Charles Lyell's seminal *Principles of Geology*, Bain took advantage of the opportunities afforded by his new profession to accumulate rock and fossil samples—the latter belonging to extinct species hitherto unknown to science. He donated samples to the Geological Society of London and also contributed a paper "On the Geology of Southern Africa" accompanied by one of the first geological maps of the region. Bain's map mirrored early European attempts. Georges Cuvier and Alexandre Brongniart drew up the first geological map of France in 1811. This was followed by the famed *A Delineation of the Strata of England and Wales, with Part of Scotland* in 1815 by English geologist, William Smith, which influenced the work of Armand Petit Defrenoy and Leonce Elie de Beaumont, mining engineers tasked with creating a similar map depicting French geology.[9] Like many of the early geologists,

[6] A. W. Rogers, *The Pioneers in South African Geology and their Work (Transactions of the Geological Society of South Africa, Annexure to Volume XXXIX* (Geological Society of South Africa, 1937), pp. 7, 15.

[7] S.H. Haughton, 'South African Geology' in *A History of Scientific Endeavour in South Africa: A collection of essays published on the occasion of the Centenary of the Royal Society of South Africa,* A.C. Brown (ed) (Cape Town, The Royal Society of South Africa, 1977), pp. 339–340.

[8] Rogers, *The Pioneers in South African Geology*, p. 19.

[9] Gohau, *A History of Geology*, pp. 136–137.

however, Andrew Geddes Bain was not formally trained in the discipline.[10] This was soon to change.

By the later nineteenth century, geology was increasingly moving away from the realm of the amateur adventurer explorer to that of the professional. An indication of this shift was the growth of formal associations. Prior to the founding of the Geological Society of South Africa in 1895, the first formal association of geologists was the short-lived South African Geological Association (SAGA) formed in 1888 and disbanded in 1890. Its origins lay very much within the context of empire and its founding was suggested by John Shaw of the South African College at a presentation at the Queen's Jubilee South African Exhibition in 1888.[11] The role of the society was the collation of geological knowledge in the form of books, papers and maps as well as the collection and identification of rock samples submitted by members around the country. Its aims were remarkably similar to that of the very first Geological Society formed in London 80 years earlier:

> That there be forthwith instituted a Geological Society for the purpose of making geologists acquainted with each other, of stimulating their zeal, of inducing them to adopt one nomenclature, of facilitating the communications of new facts and of ascertaining what is known in their science and what remains to be discovered.[12]

In addition to geology, the South African Geological Association also included within its purview the related fields of archaeology and palaeontology. SAGA was, however, to have an extremely short lifespan, possibly indicative of the close association of geology with the practical applications of mining rather than its more academic role. With its headquarters in Grahamstown, SAGA was distant from the new centres of mining, first in Kimberley and then on the Witwatersrand. It was at the

[10] Gerry Levin, 'Andrew Geddes Bain' in 'South African Geological Association (1888–1890)' reprinted in *A Century of Geological Endeavour in Southern Africa, 1895–1995*, C.R. Anhaeusser (ed) (Linden, The Geological Society of South Africa, 1997), p. 2.

[11] Cornelus Plug and Gary Levin, 'South African Geological Association (1888–1890)' in *A Century of Geological Endeavour in Southern Africa, 1895–1995*, C.R. Anhaeusser (ed) (Linden, The Geological Society of South Africa, 1997), p. 8.

[12] http://www.geolsoc.org.uk/history, Accessed 12 December 2014.

hub of the embryonic gold mining industry on the Witwatersrand that the Geological Society of South Africa (GSSA) was established in 1895.[13]

Like SAGA, the formation of the GSSA was not isolated from international trends. David Draper, the son of a botanist, was born in the Cape Colony in the mid-nineteenth century. The diamond discoveries at Kimberley inspired his career as a geologist. Like many of the geological pioneers of South Africa, his training was not formalised and was instead based on practical experience at the diamond, gold and coalfields scattered across the country. Draper was the first South African geologist to become a Fellow of the Royal Geological Society of London and published a series of papers with that institution on the geology of South Africa. He subsequently concluded that a similar organisation would be of tremendous value in the South African context as the substantial diamond and gold discoveries led to a scouring of the country in search of new sources of mineral wealth. Thus, inspired by a British organisation, imperial networks of knowledge and the drive for the extraction of natural mineral resources the Geological Society of South Africa came into existence in 1895.[14] The same year also saw the establishment of the Cape of Good Hope Geological Survey, continuing the formalisation of geology as an established science in the country.[15] While these geological associations were inspired by similar European and American institutions, they were also driven by local concerns. A mining industry that was increasingly capital and labour intensive and had begun to dominate the economic and political structures of South Africa left little place for the adventurer prospector operating on a "hunch", albeit an informed one.

Yet the discipline of geology in conjunction with the prospect of material reward meant that the path to professionalisation was somewhat rocky. Described as "the Wizard Geologist", Hans Merensky was born in the Transvaal in 1871, the son of German missionaries. He spent much of his youth in Germany where he studied both geology and mining and returned to South Africa at the turn of the century. This was the period of

[13] Plug and Levin, 'South African Geological Association', pp. 8, 10.

[14] 'Dr David Draper (1849–1929)' in *A Century of Geological Endeavour in Southern Africa, 1895–1995*, C.R. Anhaeusser (ed) (Linden, The Geological Society of South Africa, 1997), p. 24.

[15] S. Meiring Naude and A.C. Brown, 'The Growth of Scientific Institutions in South Africa' in *A History of Scientific Endeavour in South Africa: A collection of essays published on the occasion of the Centenary of the Royal Society of South Africa*, A.C. Brown (ed) (Cape Town, The Royal Society of South Africa, 1977), p. 70.

the "consulting geologist" upon whom fortunes were made and lost. Two types of geologists predominated on the Witwatersrand. The first was the generation of men like Bain whose expertise was based on their practical experience in the field. The second was exemplified by Merensky—the formally trained geologist *au fait* with the latest scientific trends. Merensky was to experience the disjuncture between science and wishful thinking only a year after his return to South Africa. The discovery of deposits of tin in the Transvaal fuelled speculation. Over a heady period of two months the price of tin shares rose and fell dramatically based on preliminary geological reports and rumour; speculators made millions that were just as easily lost in an instant. It was here that Merensky realised the implications of early reports based on insufficient scientific investigation.[16] It was his discovery of platinum in South Africa in 1924 that marked the triumph of the professional geologist—coming just as the virtues of science in South Africa were being expounded by Smuts and his contemporaries. At the same time, it would be misleading to dismiss experience and local understandings. Since its inception, geology—which is based in the field—has used practical knowledge derived from those with closest ties to the land. Early geological specimens were collected by those who worked the land as farmers or miners. The latter, in particular, may not have been acquainted with the latest scientific theories and were not necessarily able to theorise their findings but, nonetheless, had vital practical knowledge.[17] The disjuncture between the professional and the amateur in the field of geology is more often than not a matter of degree and the mining industry in South Africa, particularly in its formative stages, required the input of both.

Diamonds, gold, copper, tin and platinum—the geology of southern Africa created the conditions for an abundance of mineral resources that became the bedrock of the South African economy. The location and extraction of these resources in turn advanced geological science in the country. However, even more significantly, mining had socio-economic and political repercussions that were to leave an indelible mark on society. The geology of South Africa became the basis for shaping the nation.

[16] Eberhard W. Machens, *Platinum, Gold and Diamonds: The adventure of Hans Merensky's discoveries*, Idette Noome (translator) (Pretoria, Protea Book House, 2013), pp. 7, 13–14, 44–46.

[17] Martin J.S. Rudwick, *Bursting the Limits of Time: The Reconstruction of Geology in the Age of Revolution* (Chicago, University of Chicago Press, 2005), pp. 32–33.

This, then, was the world into which Alexander Logie Du Toit was born on the 14th of March in 1878 on an estate known as Klein Schuur, adjacent to the more well-known Groot Schuur with a skyline dominated by Devil's Peak and Table Mountain.[18] These iconographic South African landmarks form part of the Table Mountain Group and comprise the sedimentary rock sandstone that was largely deposited during the Palaeozoic era 450 million years ago.[19] Through his formative years, then, Du Toit would have been aware of these massive geological features that left an indelible impression on him and would come to shape much of his future career.

Du Toit's Education in South Africa

The Diocesan College was established in 1849 in Rondebosch, a suburb of Cape Town, over which looms the region's defining geographical feature—Table Mountain. Popularly known as "Bishops", it was one of the two main colleges in southern Africa—the other being the South African College. The colleges provided teaching and training, preparing students for the examinations written under the auspices of the University of the Cape of Good Hope. By the 1870s, the Diocesan College placed an importance on science education—by the time of Du Toit's birth in 1878 C. Lloyd Morgan had been employed. Reflective of the strong ties of empire, Morgan had been taken under the tutelage of ardent Darwin supporter and biologist, Thomas Henry Huxley.[20] It was here that Du Toit completed his secondary school education before enrolling at the University of the Cape of Good Hope—which would eventually become the University of Cape Town.[21]

The University of the Cape of Good Hope was based in Cape Town. The first university in southern Africa, it was established five years before

[18] Gevers, "The Life and Work of Dr. Alex. L. Du Toit", p. 1.

[19] A.G. Thamm and M.R. Johnson, "The Cape Supergroup" in *The Geology of South Africa*. M.R. Johnson, C.R. Anhaeusser and R.J. Thomas (eds) (Johannesburg and Pretoria, The Geological Society of South Africa and the Council for Geoscience), p. 444.

[20] M. Boucher, *The University of the Cape of Good Hope and the University of South Africa, 1873–1946: A Study in National and Imperial Perspective* (Pretoria, The Government Printer, 1974), pp. 25, 31, 41, 55.

[21] Gevers, "The Life and Work of Dr. Alex. L. Du Toit", p. 2.

Du Toit's birth.[22] The university symbolised two aspects of embryonic South Africa—the first was an assertion of a growing intellectual identity; the other was its position as part of the British Empire. Inspired by the University of London, "each colonial reflection of the London example was an empire in microcosm".[23] The competing pull between former colony and metropole, independence and interdependence would contextualise Du Toit's geological research. For now, though Du Toit was a student and graduated with a BA Honours degree and, in the vein of many middle class white South Africans, went abroad for further study. His destination was Scotland—his ancestral homeland and the birthplace of modern geology.[24]

Scotland

After graduating from the University of the Cape of Good Hope, Du Toit enrolled at the Royal Technical College at Glasgow, completing a degree in Mining Engineering in 1899 which he then followed up with further geological studies at London's Royal College of Science before returning to Scotland.[25] As a student, Du Toit demonstrated the physical fitness and love for active geological exploration that would remain with him for the rest of his life. As a member of a cycling club he would explore the area around Glasgow, acquainting himself with its geology and even made trips to Edinburgh—the birthplace of modern geology—to peruse the geological specimens housed at the Edinburgh Museum and scrutinise the maps at the Geological Survey.[26]

His student years in Glasgow were also memorable for Du Toit meeting and marrying Adelaide Walker who would return with him to South Africa. Their marriage would produce a son, Alexander Robert du Toit; both mother and son accompanied Du Toit on his early geological expeditions.[27]

[22] Howard Phillips, *The University of Cape Town, 1918–1948: The Formative Years* (Cape Town, University of Cape Town Press, 1993), p. 1.

[23] Boucher, *The University of the Cape of Good Hope and the University of South Africa*, p. 23.

[24] Gevers, "The Life and Work of Dr. Alex. L. Du Toit", p. 2.

[25] Gevers, "The Life and Work of Dr. Alex. L. Du Toit", p. 2.

[26] University of Cape Town Libraries—Jagger Library (hereafter UCT-JL): Alex L. Du Toit Papers—BC 722: A4—Personal Documents: rough biographical notes.

[27] Gevers, "The Life and Work of Dr. Alex. L. Du Toit", p. 2.

The Glasgow and West of Scotland Technical College developed out of the merging of Anderson College—devoted to technical education and named after John Anderson, who opposed Hutton's uniformitarianism in favour of Biblical-influenced catastrophism[28]—and the College of Science and Arts, amongst others, in 1887. By 1912 the institution was known as the Royal College of Technology and, a year later, was officially associated with Glasgow University.[29]

Du Toit's early professional career pre-empted this association. It was upon completing his studies at the Royal College of Science in 1901 that Du Toit—then only 23 years old—was offered the position of Lecturer in Surveying, Mining and Geology at his alma mater, the Royal Technical College in Glasgow as well as Lecturer in Geology at the University of Glasgow—the latter having a particular prestigious association with the development of modern scientific thought.[30]

While Edinburgh remained the focus of the Scottish Enlightenment, it was Glasgow—and particularly the University of Glasgow—that was the centre of science or "natural philosophy" in the eighteenth century. Its alumni included Adam Smith, William Cullen and his successor in chemistry, Joseph Black. Influenced by the pioneer in chemistry Antoine Lavoisier, Black focused his research on heat—with the aid of the instruments made for him by James Watt—which, as mentioned previously, would go on to have a dramatic impact on Hutton's own theories of the heat in the earth's interior.[31]

In the mid-nineteenth century, concomitant with the University's attempt to maintain its teaching philosophy of exposing an Honours candidate to a range of subjects and with its continuing emphasis on science, the department of Natural Science came into being (along with the existing departments of Mathematics and Natural Philosophy, Classics and Philosophy). This new department would comprise Chemistry, Zoology, Botany and Geology.[32] In 1873, the Bachelor of Science degree was initiated and a Faculty of Science was formed in 1893, shortly before

[28] David B. Wilson, *Seeking Nature's Logic: Natural Philosophy in the Scottish Enlightenment* (University Park, Pennsylvania State University Press, 2009), p. 196.

[29] R. D. Anderson, *Education and the Scottish People, 1750–1918* (Oxford, Oxford University Press, 1995), pp. 269, 277.

[30] Gevers, "The Life and Work of Dr. Alex. L. Du Toit", p. 2.

[31] Wilson, *Seeking Nature's Logic*, pp. 72–73, 133, 135.

[32] J.D. Mackie, *The University of Glasgow, 1451–1951: A Short History* (Glasgow, Jackson, Son and Company, 1954), pp. 273, 279–280.

Du Toit's arrival. The University offered the Doctorate in Science in 1908[33] and Du Toit was able to capitalise on this, when he was awarded his D.Sc. in 1910.

Early Career in South Africa

Du Toit's stint in Scotland had meant that he had not been a part of one of the defining aspects of creating South Africa—the South African War, which had pitted Boer against Briton and had significant implications for the association between the nation and science. The Geological Society of South Africa was founded in 1895 by David Draper. Its establishment reflected a convergence of international and local trends in the creation of scientific knowledge. Science in late nineteenth and early twentieth century South Africa could not be divorced from the political, social and intellectual aspirations of its setter population. This was a pivotal period in the nascent South Africa, marking the end of three years of bitter enmity between Britain and the former Boer republics that had drawn in other groups as well.[34] The year 1910 saw the creation of the Union of South Africa, a union of two former British colonies and two former Boer republics. The term "union" is something of a misnomer as it papered over the tensions that existed in South African society between Boer and Briton, black and white. As it had become an assertion of a particular form of settler identity asserting itself against British imperialism while, nevertheless, incorporating British knowledge systems in the nineteenth century, science now became a means in forging a national identity from disparate elements. The year 1910 was also the year that the Geological Survey of South Africa was established.[35] The measurement, control and regulation of land and the acquisition of knowledge regarding its mineral wealth were necessary components in forging a nation that was to be based on inequality, exclusion and dispossession.

When Du Toit returned to South Africa in 1903, he would contribute to this assertion of national identity. From this period, Du Toit would maintain a diary for every year. The diaries kept by Du Toit in the early

[33] Mackie, *The University of Glasgow*, p. 297.

[34] Shula Marks, "War and Union, 1899–1910" in *The Cambridge History of South Africa, Volume 2: 1885–1994*, R. Ross, A.K. Melger and B Nasson (eds) (Cambridge, Cambridge University Press, 2011), p. 163.

[35] Naude and Brown, 'The Growth of Scientific Institutions in South Africa', p. 71.

decades of the twentieth century were produced by Lett's, each one retailing for a shilling, with the year printed on the cover and are described either as "South African Rough Diary" or—as in the case for the one penned in 1914—"South African Union Scribbling Diary". Each page is filled with brief entries penned in small handwriting (however, writing style and size can vary, indicative of writing under different conditions, symbolic of the vagaries of geological exploration).[36]

Du Toit became a member of the Geological Commission of the Cape of Good Hope in 1903 and the maintenance of a diary from this period forward is perhaps an indication of the significance of the event—the beginning of a long professional career that would prove to be both illustrious and controversial. The diary for 1903 followed a different format from later diaries with more detailed entries that would later become formulaic and spare. Written in a plain notebook, it begins with heading "Diary 1903" followed by "Appointed Assistant-Geologist". It reflects the enthusiasm of a 25-year-old appointed in his first position as a member of the Geological Commission.[37]

Also formed in 1895 as part of the Department of Agriculture, the Geological Commission was responsible for the surveying and drawing up of geological maps. The chairman of the Geological Commission was leading Cape politician John Xavier Merriman and his report, citing Du Toit's new appointment, stated that the position of geologist, "'been filled by a **colonist**, Mr A.L. Du Toit, who after a distinguished college career in South Africa, qualified in Europe both in theoretical and practical work... He promises to be a valuable addition to the staff'".[38] The emphasis on "colonist" has most likely been added by Gevers in his biography of Du Toit and Gevers takes particular exception to what he perceives to be the slighting description of Du Toit as a mere colonist:

> Little did John X. Merriman, that well-known political figure at the Old Cape, realise, I am sure how vastly his expectations would be exceeded and that the name of the young **colonist** would one day become known far beyond the boundaries of the Old Colony and achieve world fame among scientific circles right around the globe.[39]

[36] UCT-JL: Alex L. Du Toit Papers—BC 722: A1—Diaries.
[37] UCT-JL: Alex L. Du Toit Papers—BC 722: A1—Diaries: 1903.
[38] Gevers, "The Life and Work of Dr. Alex. L. Du Toit", p. 3.
[39] Gevers, "The Life and Work of Dr. Alex. L. Du Toit", p. 3.

Yet Merriman's espousal of Du Toit as a "colonist" also spoke to the politician's own liberal, anti-imperial leanings. Merriman was born in Somerset, England in 1841 and his family left England for South Africa in 1848 so that his father, Nathaniel, could take on the role of archdeacon of Grahamstown. After rather desultory attempts at employment as a young man—including a stint as a surveyor—Merriman found his true calling in politics, as a Cape liberal. In the wake of the South African War, he was unequivocal in his desire for an independent South Africa free from imperial domination, promoting the liberal tradition associated with the Cape.[40] Du Toit's appointment to the Geological Commission can therefore be placed within this context of liberalism and incipient nationhood.

A.W. Rogers was the Acting Geologist for the Geological Commission. He had been working alongside E.H.L. Schwarz, mapping the Cape, until the latter took the post of Professor of Geology at Rhodes University. After Schwarz's departure, it was left to Alex Du Toit who, under the tutelage of Rogers, was assigned to survey and ultimately map a vast area of almost 420,000 square kilometres. Rogers was the ideal mentor for the young Du Toit, emphasising inductive reasoning and the importance of fieldwork. According to Gevers, Du Toit was also physically and mentally equipped for the mapping of vast terrains and its concomitant discomforts, being "spare of frame and abstemious in his habits, adventurous and full of vigour… His body was as active and agile as his mind". And this would serve him in good stead. For the next seven years from 1905 until 1912, Du Toit and Rogers worked on the Karoo System, with particular focus on the Dwyka Tillite. The observations that Du Toit made here would underpin his later advocacy of continental drift and also laid the foundations for a lifelong interest—and expertise—in volcanism and glaciation.[41]

Evident, too, in his early fieldwork was Du Toit's emphasis on the fossil record with his reports highlighting the faunal and floral fossil evidence found in the Karoo stratigraphic succession. He consulted with fossil experts, Alfred Brown and D.V. Kannemeyer in his survey of the Stormberg Series and Beaufort Series. The resulting report—with accompanying map—totalled more than a hundred pages and included an account of the fossilised reptiles, amphibians and flora that distinguished the strata. Also

[40] Phyllis Lewsen, *John X. Merriman: Paradoxical South African Statesman* (Johannesburg, AD Donker, 1982), pp. 2–4, 15, 254.

[41] Gevers, "The Life and Work of Dr. Alex. L. Du Toit", pp. 3–5.

considered were the samples of fossilised wood and the preservation of insect remains in the sandstone strata termed the Red Beds. His early studies of the *Thinnfeldia* flora, characteristic of the Molteno Beds, prompted him to write "The Fossil Flora of the Upper Karoo Beds" in 1927, which established him as a pre-eminent palaeobotanist. Of particular interest as well was Du Toit's account of a region known as "Fossil Hill", from which would be unearthed the fossilised remains of the hominin, *Australopithecus africanus* decades later.[42] In March of 1903 Du Toit discovered Glossopteris fossils as he was examining rocks alongside a railway. This fossilised fern would soon assume great significance.[43]

Du Toit's particular aptitude lay in mapping, a talent for which he was held in some esteem by his peers—as well as succeeding generations of geologists. As the name implies, geological mapping involves the tracking of geological features that are evident at the surface and can contain a wealth of information. These maps show the visible strata as well as what the geologist is able to infer lies beneath the earth's surface and describe the different types of rock formations and layers and their composition as well as their origin in terms of geological processes such as volcanism, for instance. Also included are sites of potential resources such as coal or water as well as possible geological hazards.[44] In his very first year as a field geologist, Du Toit was able to map the Cape Peninsula which took him a mere three weeks. It was an early indication of his talent for constructing a whole out of disparate elements, the ability to see the entire picture. Admiring accounts were told of Du Toit's discontent with the existing maps with which he was supplied, compelling him to effortlessly draw up his own in the field.[45]

A large portion of Du Toit's fieldwork had also involved the Karoo System, the largest in South Africa, which encompassed much of the country. Du Toit was the world's leading authority on this system—composed largely of sedimentary rock—and he would follow its sequence

[42] Gevers, "The Life and Work of Dr. Alex. L. Du Toit", pp. 6, 9, 13, 106.

[43] UCT-JL: Alex L. Du Toit Papers—BC 722: A1—Diaries: Saturday, 21 March 1903.

[44] Broken Hill—Olary Geology and Geophysics Field Course, http://www.geosci.usyd.edu.au/users/prey/FieldTrips/BrokenHillOlary/Mapping.html, accessed 17 January 2016.

[45] Gevers, "The Life and Work of Dr. Alex. L. Du Toit", pp. 6, 8.

to the landmasses that once comprised Gondwana. His report on the copper and nickel-bearing Insizwa Range of the Karoo System highlighted yet another facet of his career as a geologist—petrology and mineralogy. In an era of growing specialisation, this was also a demonstration of Du Toit's enviable ability to drawn upon the different subfields of geology in order to construct a comprehensive picture of the earth's history. As a result, he was subsequently awarded his D.Sc. or doctorate in science from the University of Glasgow in 1910.[46]

Geological mapping in difficult, frequently impassable terrain in sometimes adverse climatic conditions demanded the resilience and resolution that Du Toit was beginning to demonstrate very early on. Roads were often in poor condition—if not non-existent—and Du Toit travelled on a wagon pulled by a donkey or, in even more inaccessible areas, on a pushbike. In writing his appreciation of Du Toit, Gevers was regaled with stories of his exploits:

> Many a farmer in the Stormberg and trader in Pondoland delighted me with reminiscences of Du Toit. Slinging rucksack, plane table and tripod over his shoulder, rushing around on his bicycle in country least suited for such locomotion, fording streams with bicycle and all on his back, climbing every mountain top and scrambling down into every ravine and donga in all seasons of the year, he had undoubtedly impressed them as a very demon for work and indefatigable energy.[47]

[46] Gevers, "The Life and Work of Dr. Alex. L. Du Toit", pp. 9–10, 15.
[47] Gevers, "The Life and Work of Dr. Alex. L. Du Toit", pp. 7–8.

Traversing difficult terrain: Alex Du Toit on horseback (University of Cape Town Libraries, mss_bc722_q2a)

Entries made in the very first year of his career are a sobering illustration of geological fieldwork in remote areas that could only be accessed on foot: "Ascended remainder of Verlaten Kloof; waggon [sic] stuck ¼ mile from top and had to be drawn up by 2 teams of donkeys".[48] During the course of his mapping an unfortunate donkey was injured and later died. Difficulty in travelling was exacerbated by adverse weather, especially the cold, windy conditions of winter.[49]

An early photograph of Du Toit in the field evokes the pioneering days of a bygone era. The covered cart is in the background, the words "Geological Commission Cape Town" emblazoned on the side. A horse grazes in the right foreground. Seated in front of the cart is a young, smiling woman, with a little boy beside her. This was Du Toit's wife, Adelaide, who he married whilst studying in Glasgow and their young son, Robert. A moustachioed Du Toit sits a little apart from them.[50] The figure portrayed is of the dedicated geologist, the rugged adventurer in the service of science, braving adversity in order to name and map, and a descendent of the hardy settlers who had tamed a hostile land.

The bulk of Du Toit's fieldwork was conducted over almost two decades from 1903 to 1920 with a brief hiatus during a trip to Australia and the First World War. During this period he traversed and mapped more than 80,000 square kilometres of South Africa. The first geological map of the new country was issued in 1926, with several of its boundaries based upon Du Toit's fieldwork. By 1922, Canadian geologist R.A. Daly of Harvard University—and an early proponent of continental drift theory—would describe Du Toit as the "world's greatest field geologist", an appellation that would eventually set him on the path to South America.[51]

[48] UCT-JL: Alex L. Du Toit Papers—BC 722: A1—Diaries: Saturday, 21 March 1903.
[49] UCT-JL: Alex L. Du Toit Papers—BC 722: A1—Diaries: 1903.
[50] Gevers, "The Life and Work of Dr. Alex. L. Du Toit", p. 2 and Photo 2.
[51] Gevers, "The Life and Work of Dr. Alex. L. Du Toit", pp. 8, 24.

CHAPTER 3

A World in a Grain of Sand: A Brief History of Geology and the Origins of Continental Drift Theory

Alex Du Toit's education in Scotland would have placed him firmly in the uniformitarianism of Scottish geology's founding fathers, James Hutton and Charles Lyell. Yet his advocacy of continental drift would also challenge what had increasingly become dogma and further highlight the way in which the history of geology has been defined by controversy. In *Time's Arrow, Time's Cycle*, Stephen Jay Gould addresses the influence of "whiggish history" or a triumphalist tale of progress in science which frees science from the irrational and supernatural past.[1] The history of geology and of continental drift challenges this in two ways. James Hutton's uniformitarianism did not present an absolute break with religion. Moreover, as uniformitarianism became the fundamental means of understanding geological processes, it became dogma, presenting a key obstacle to the reception of the theory of continental drift.

The Scientific Revolution valourised scientific endeavour as the means of shining rationality and reason into the dark corners of superstition. In the words of Carl Sagan, science was a "candle in the dark".[2] Stephen Jay

The title of this chapter is taken from William Blake "Augeries of Innocence".

[1] Stephen Jay Gould, *Time's Arrow, Time's Cycle: Myth and Metaphor in the Discovery of Geological Time.* (Penguin Books, 1991) pp. 4–5.

[2] Carl Sagan, *The Demon-Haunted World: Science as a Candle in the Dark.* (New York, Ballantine Books, 1997).

S. Chetty, *'Africa Forms the Key'*, Cambridge Imperial and Post-Colonial Studies, https://doi.org/10.1007/978-3-030-52711-2_3

Gould uses the Burgess Shale to demonstrate the way in which science—and scientists—cannot be divorced from their context. A myriad of invertebrate fossils dating from the Cambrian explosion—the first flourishing of complex life—was discovered in the Canadian Rockies by geologist, Charles Doolittle Walcott in 1909.[3] Working with his family, Walcott obtained thousands of specimens of creatures never before seen from the Burgess Shale, a collapsed prehistoric mud bed. An unspoken assumption of Darwinism, evident in the pictorial depictions of human evolution and reminiscent of the Great Chain of Being, is that life evolves from the simple to the complex—with modern human beings at the apex of the evolutionary tree. The conflation of evolutionary change with progress was something Darwin himself found difficult to resolve. His theory of natural selection posited that, within natural variance, the life forms most likely to survive would be those most suited to a particular environment at a particular point in time. Yet Darwin was himself a product of the Victorian era with its unshakeable and optimistic belief in improvement. In this way, the description of natural selection became imbued with a value judgement—progress. Walcott, a religious man and a Darwinist, interpreted his findings within this world view. The fantastical and complex creatures within the Burgess Shale were the antecedents of modern life forms.[4]

Walcott's interpretation remained unchallenged for decades until an expedition to the Burgess Shale was mounted under the leadership of palaeontologist Harry Whittington and geologist J.D. Aitken in 1966 and 1967. Whittington's reassessment of Walcott's samples in particular was a revelation. His exhaustive study of a number of the fossil finds provided overwhelming evidence that they could not be neatly situated within existing taxonomies—they were examples of life forms hitherto unknown. Many of these life forms subsequently went extinct and life as we know it is derived from the few that remained—and is therefore less varied. Moreover, the detailed examination of the Burgess Shale fauna suggested no reason for their extinction—these life forms seemed no less suited for survival than our own ancestors. This left one with the discomfiting conclusion that extinction and survival may be somewhat arbitrary.[5] The story

[3] Stephen Jay Gould, *Wonderful Life: The Burgess Shale and the Nature of History*. (New York, WW Norton and Company, 1990) pp. 23–24.

[4] Gould, *Wonderful Life*, pp. 69, 28–35, 257–258, 46.

[5] Gould, *Wonderful Life*, pp. 77, 123–124, 79, 208, 233–239.

of Charles Doolittle Walcott and the Burgess Shale is a narrative of the ideological constraints within which science operates and the assumptions that underlie scientific claims to truth. The reinterpretation of the fossils of the Burgess Shale undermines the belief in the value of our own existence which was something that Walcott was unwilling—or, more likely, unable—to do.

Gould's account of the Burgess Shale and the presumptions of Walcott demonstrate the lack of linear progression of science and truth. This is reinforced by the development of geology as a discipline from the nineteenth century.

Modern geology can be divided into two phases. The first has its origins in Scotland in the work of Charles Lyell and his predecessor, James Hutton. The second began in the twentieth century with plate tectonics. As Du Toit began his education in Glasgow he was, in a sense, straddling these two worlds.[6]

James Hutton was a product of the Scottish Enlightenment. He would complete his medical training at the University of Paris and the University of Leyden where his thesis on the circulatory system was modelled on Newton's view of orbiting planets which he would subsequently apply to the earth itself.[7] At this point, geology was an embryonic science and the few texts dealing with the subject sought to reconcile geology with biblical understandings of the earth's creation. The exception was the *Histoire Naturelle* by Comte de Buffon which put forward the heretical view that landforms were a result of the recession of the ocean rather than the direct hand of God. Hutton began developing his own understanding of the earth's mechanisms and, unlike the other geologists of his time, he was able to bring his own scientific training in chemistry to bear on his views. His first published foray into geology came as a result of his farming activities where he proposed that rocks were sedimentary in nature, formed by the processes of erosion. While hardly novel, the implications suggested a rock cycle that was in perpetual motion and would have further implications for the age of the earth.[8]

[6] Preston Cloud, "The Second Flowering of Geology" in *Bulletin of the American Academy of Arts and Sciences*, Vol. 36, No. 3 (Dec, 1982) pp. 34–36.

[7] Jack Repcheck, *The Man Who Found Time: James Hutton and the Discovery of the Earth's Anatomy*. (New York, Basic Books, 2009) pp. 45–494, 88–92, 58–59, 60, 6.

[8] Repcheck, *The Man Who Found Time*, pp. 91–100, 103, 113–114.

Hutton became particularly friendly with pioneer chemist, Joseph Black, due to their mutual interest in chemistry. Moreover, Black's work on heat and pressure would influence Hutton's views of the interior of the earth. Both men collaborated in the field of mineralogy. Conventional wisdom suggested that minerals had precipitated in the ocean; however, Hutton and Black suggested that it was heat from the earth's interior, in conjunction with pressure, that was responsible for the process of mineralisation.[9]

Through his association with Black, Hutton was introduced to Enlightenment luminaries James Watt and Adam Smith and, along with the latter, he and Black would go on to form the Oyster Club. Hutton was also a member of the Philosophical Society and its successor, the Royal Society of Edinburgh. It was to the Royal Society that he was asked to present his work in March, 1785. The controversy came in his second lecture delivered a month later where Hutton proposed that rock strata were uplifted, not by receding water which would leave the layers of sedimentary rock perfectly horizontal, but by the internal heat of the earth which created melted rock and was responsible for the distortion so clearly visible in rock outcrops. The corollary of this was the earth was incredibly old to have sustained successive cycles of erosion, sedimentation and uplift.[10]

Roy Porter argues that Hutton's vision of the geological processes that shaped the earth was shaped by the Scottish Enlightenment and "its philosophical grasp of reality" as evident in the work of Adam Smith. The human senses perceived information thus all data was mediated by the mind. This then questioned the status of the pure, empirical "fact" evident in Francis Bacon's model. All sensory input was assessed and incorporated into a form of understanding, a means of bringing "order" and structure to the natural world. Hutton's observations of geological processes would therefore lead to an understanding of geological processes across time and space. And, challenging the break with religion, Porter further argues that the highest understanding of "natural philosophy" was "natural religion". Behind Hutton's geological model was that the will to understand nature was to understand the work of God and that a view of the earth's processes

[9] Repcheck, *The Man Who Found Time*, pp. 117, 120, 129, 132–134.
[10] Repcheck, *The Man Who Found Time*, pp. 134, 140–145, 151–153.

as constant—without capricious cataclysm—symbolised the perfection of the divine.[11]

In addition to a philosophical perspective, Hutton's understanding was based on the evidence of existing rock formations such as Siccar Point, for instance, consisting of a sedimentary rock formation where the lower strata is tilted until it is almost vertical. This indicated an almost unimaginable time period to allow for these slow geological processes to occur.[12] Confronted with Siccar Point, Hutton's companion, John Playfair, articulated its significance, "The mind seemed to grow giddy by looking so far into the abyss of time".[13] Hutton's findings were published in the seminal—but, by all accounts, largely unreadable—*The Theory of the Earth* in 1795, two years before his death.[14] His findings, however, would usher in a new understanding of geology—uniformitarianism—and find their champion in the man who became the face of modern geology, Charles Lyell.

The history of modern geology can be narrated as a series of episodes of dissent and debate—the formulation of a new theory, the disagreements—sometimes acrimonious—between its opponents and proponents and, finally, provided sufficient supporting evidence has been unearthed, its acceptance into mainstream thought. In addition, each controversy often carries with it the seeds of earlier conflict—knowledge is not created in a vacuum. Geology in the eighteenth century was defined by two dominant schools of thought—the neptunists, exemplified by Abraham Werner, who proposed that all geological features on the surface of the earth were as a result of the action of water and "chemical precipitation from the universal ocean". The plutonists, on the other hand—into whose camp, as we have seen, James Hutton fell—proposed a more dynamic view of the earth with geological processes being driven by heat from the earth's interior which uplifted strata, subjecting them to erosion and subsequent deposition of eroded material in the oceans, thus creating the conditions for the formation of new strata.[15]

[11] Roy Porter, *The Making of Geology: Earth Science in Britain, 1660–1815*. (Cambridge, Cambridge University Press, 1977) pp. 190–192.

[12] A. Hallam, *Great Geological Controversies*. (Oxford, Oxford University Press, 1992) pp. 36–38, 41–43, 34, 31–33.

[13] Quoted in Richard Fortey, *The Hidden Landscape: A Journey into the Geological Past*. (London, Bodley Head, 2010) p. 68.

[14] Repcheck, *The Man Who Found Time*, pp. 58–59. pp. 151–158, 161.

[15] A. Hallam, *Great Geological Controversies*, pp. 5, 11–13.

Hutton's uniformitarianism would be inherited and refined by Lyell. While initially indifferent to the scientific advances made throughout Europe, the period just prior to Lyell's entry into Oxford University was marked by change, particularly evident in chemistry and mineralogy with the latter under the leadership of William Buckland.[16]

An expert in fossils and a reverend, Buckland embraced catastrophism as it reconciled geology with his own religious beliefs. Influenced by Cuvier, he interpreted the fossilised remains of extinct creatures found in a Yorkshire cave as well as the fossils of marine creatures discovered on high mountain ranges as being the products of a great flood or deluge—as described in the Old Testament.[17] By all accounts, Buckland's lectures on mineralogy and his expertise in fossils whet the appetite of young Lyell, awakening in him an interest in geology that would have far-reaching implications for the discipline. Lyell completed his degree at Oxford in 1819 and, in that same year, was made a fellow of the Geological Society of London as well as the Linnean Society.[18]

In 1824, Lyell was taken to Siccar Point, site of Hutton's own insights and this encounter contributed to Lyell's eventual espousal of uniformitarianism.[19] Four years later, Lyell travelled to Auvergne with fellow Geological Society member, Roderick Impey Murchison—also a catastrophist. This proved to be one of Lyell's earliest exposures to geology in the field and he was particularly taken with the layers of sedimentary rock accumulated throughout a long period of deposition. Each layer was incredibly thin but the total accumulation was approximately 230 metres high. For Lyell, this indicated a long and uninterrupted period of deposition that he estimated to have taken hundreds of millennia.[20] Ironically, despite the early influences of catastrophists Buckland and Murchison, Lyell's deduction at Auvergne would make him the strongest proponent of uniformitarianism.

There were two key features of uniformitarianism. The first was continuity—all geological change was slow and continuous. Through these

[16] Leonard G. Wilson, *Charles Lyell: The Years to 1841: The Revolution in Geology*. (New York and London, Yale University Press, 1972) pp. 33, 34.

[17] Martin J.S. Rudwick, *The Meaning of Fossils: Episodes in the History of Palaeontology*. (New York, Science History Publications, 1976) pp. 135–136.

[18] Wilson, *Charles Lyell*, pp. 44, 64.

[19] Repcheck, *The Man Who Found Time*, p. 179.

[20] Gabriel Gohau, *A History of Geology*, Albert V. Carozzi and Marguerite Carozzi, (translators). (New Brunswick and London, Rutgers University Press, 1990) p. 140.

incremental changes great transformations of the landscape were possible, discounting any need for catastrophes. The other aspect was what has been termed a "steady state model".[21] Geological change was constant—observable processes in the present mirrored those in the past and were applicable to the entire globe. This inevitably brought uniformitarianism into conflict with catastrophism which equated change with linear progression—geological change contributed to a process of development of the Earth.[22] Ultimately, Lyell took uniformitarianism even further than Hutton, proposing that the observable geological processes of the present were, not only the same as had occurred in the past, but had always operated at a constant degree and strength. As Lyell famously put it, "The present is the key to the past".[23]

Lyell published the first of his three volumes entitled *Principles of Geology* in 1830. Its full title *Principles of Geology: An attempt to explain the former changes of the Earth's surface by reference to causes now in operation* summarised his view on uniformitarianism and his ambition to rewrite contemporary views of the history of the Earth.[24] A year later, 22-year-old Charles Darwin took with him a copy of *Principles of Geology* as he began his voyage on the *Beagle* which would prove fundamental to his own understanding of the evolution of life.[25] On reaching South America Darwin interpreted the discovery of shells raised far above sea level as evidence of uniformitarianism in action—the slow uplift of the Andes mountain chain. This was confirmed by an earthquake in 1835 where Darwin witnessed further raising of the coast. Similarly, coral reefs were indicative of subsidence, albeit at different levels of progression. Where Darwin did differ with Lyell, however, was in his belief that the evolution of living organisms did suggest a move towards greater complexity.[26]

Lyell was inspired by Darwin's accounts of his travels and, upon the latter's return to England, initiated a meeting and the beginning of a

[21] This is accredited to Martin Rudwick in Gohau, *A History of Geology*, p. 142.

[22] Gohau, *A History of Geology*, p. 142.

[23] Lisa Yount, *Alfred Wegener: Creator of the Continental Drift Theory*. (New York, Chelsea House, 2009) p. 10.

[24] Martin J.S. Rudwick, *Worlds before Adam: The Reconstruction of Geohistory in the Age of Reform*. (Chicago and London, The University of Chicago Press, 2010) pp. 297–298.

[25] Gohau, *A History of Geology*, p. 139.

[26] Rudwick, *The Meaning of Fossils*, pp. 188–189.

friendship, becoming a mentor to the younger man.[27] This friendship between Lyell and Darwin epitomised the symbiotic relationship between geology and palaeontology. Geology provided the time necessary for evolution to occur, changes that were reflected in the fossil record.[28] Fossils provided—and continue to provide—a means of dating and correlating rock strata. The conjoined nature of these disciplines would take on a special significance in the Union of South Africa where palaeoanthropology would argue for Africa's centrality in the origins of humankind and geology—in the form of continental drift—would simultaneously make a claim for the continent's geographic centrality, all of which would challenge existing belief, provoking further controversy and highlighting the relationship between science and ideology.

According to Hallam, the importance of paying particular attention to the contestations and conflict in the history of geology is that it shows the development of the discipline, complicating the notion of linear progression; further, it provides the context in which new perspectives are introduced, addressed and finally gain acceptance.[29] And, as evident by Stephen Jay Gould's discussion of the Burgess Shale, also highlight the ways in which older—and sometimes discarded—views nonetheless shape contemporary constructions of knowledge.[30] Heir to the Scottish tradition of geology of Hutton and Lyell, Alexander Du Toit would find himself embroiled in the key geological controversy of the twentieth century—continental drift.

The origin of continental drift theory is hazy—any attempt to grasp it leads to it receding further into the distant past. Before the 1960s when technology and a convergence of evidence confirmed plate tectonics—leading to its widespread acceptance—the reception of drift theory was contextualised by a mixture of myth, ideology and inference. Prior to the work of the seminal figures of drift theory—Suess and Wegener—the Blanford brothers are an early example of the complexity that underlay the understanding of the Earth's processes. Their work in the nineteenth century demonstrates the ways in which explanations for these processes were shaped by the spaces in which observations were made. Europe's

[27] Adrian Desmond and James Moore, *Darwin*. (London, Penguin Books, 2009) pp. 201–202.

[28] Repcheck, *The Man Who Found Time*, p. 194.

[29] Hallam, *Great Geological Controversies*, pvii.

[30] *Cf.* Stephen Jay Gould, *Wonderful Life: The Burgess Shale and the Nature of History*. (New York, WW Norton and Company, 1990).

topography gave a particular slant to the understanding of the geological past and it was the geology of Scotland that would usher in uniformitarianism. Travelling further afield, however, led to an early challenge to uniformitarianism and, with it, a tentative theorisation of continental drift.

Trained as geologists, William Thomas and Henry Francis Blanford left for India in the mid-nineteenth century to take up posts in the Indian Geological Survey. This period was one of increasing scientific investigation and the acquisition of knowledge of the far-flung parts of the British Empire. The work of geologists, biologists, geographers and palaeontologists began revealing some cryptic discrepancies. Species that shared a great deal of biological congruence were apparently separated by almost insurmountable distances while, conversely, those which had little in common were found in close proximity.[31]

The Blanfords were sent to survey the Talcher coalfield in the Indian state of Orissa. The coal deposits were a product of an ancient tropical environment that had fostered extensive vegetation. At the bottom of this stratigraphic succession, above bedrock that had suffered erosion, lay a deposit of unsorted material—boulders of different and fairly colossal size surrounded by fine-grained mudstone. The agents of erosion are wind, water and ice and only glacial ice has the carrying capacity to indiscriminately transport material of such varying size. The Blanfords came to only one conclusion which was the subject of papers presented before the Geological Society of London in 1877 and 1886 respectively—this humid area of the Indian subcontinent had once been covered by a glacier. Henry Blanford had knowledge of a similar stratum in South Africa, later termed the Dwyka Formation (which Du Toit would use extensively in his own research). Their superior, the Superintendent of the Geological Survey of India, Dr Thomas Oldham had noticed the same in Australia. All of this suggested extensive glaciation that seemed to have occurred at the equator at some point in the ancient past.[32] Prior to their presentation before the Geological Society, Henry Blanford had also noted the similarities in fossilised flora and fauna that would strengthen his view of an hypothesised "Indo-Oceanic" continent, "The affinities between the fossils both animals and plants, of the Beaufort group of Africa and those of the Indian Panchéts and Kámthis are such as to suggest a former existence of a land

[31] Ted Nield, *Supercontinent: Ten Billion Years in the Life of Our Planet*. (London, Granta Books, 2008) pp. 30–31.

[32] Nield, *Supercontinent*, pp. 33–34.

connexion between the two areas". This fossil similarity was also detected in comparable strata in Australia.[33]

William Blanford had temporarily left his Indian assignment to serve as a geologist under General Robert Napier in Abyssinia (present-day Ethiopia). Using the somewhat limited opportunity for fieldwork, Blanford demonstrated the same breadth of vision in detecting stratigraphic correlations across continents that would later characterise Du Toit's work. Blanford's examination of the strata led to criticism of the prevailing uniformitarian view of a steady state of erosion—a view that was based on the more moderate rate of rainfall in Europe and one that did not account for cataclysmic change that could be wrought by the action of water. Blanford's interest was also caught by the evidence of volcanic activity in the northern part of the region that seemed to bear remarkable similarity to observations that he had made in Aden, suggesting that it was a single volcanic event.[34] He would also note similar volcanic features in India, posing the question, "what connexion exists between…the lower tertiary traps of South-western Asia and Eastern Africa? Should they be proved to have been formerly connected, and to be portions of the same great volcanic origin, an idea which seems by no means improbable, their study will become one of very great interest as connected with the geological history of the earth's surface".[35] A few years later, his younger brother, Henry, travelled through Persia and would echo William's view. Henry Blanford's encounter with the extensive Zagros mountain range led to speculation of a widespread volcanic event in the Caucasus, India and Persia that had created it.[36]

In 1890 in his capacity as president of the Geological Society of London, William Blanford delivered a speech which criticised the prevailing view of the permanence of ocean basins and emphasised the importance of similar fauna in widely separated regions. The main thrust of his speech, however,

[33] Alan E. Leviton and Michele L. Aldrich, "The Impact of Travels on Scientific Knowledge: William Thomas Blanford, Henry Francis Blanford, and the Geological Survey of India, 1851–1889" in *Proceedings of the California Academy of Sciences*, Vol. 55, Supplement II, No. 9, November 19, 2004, pp. 121–122.

[34] Leviton and Aldrich, "The Impact of Travels on Scientific Knowledge", p. 122.

[35] Leviton and Aldrich, "The Impact of Travels on Scientific Knowledge", p. 125.

[36] Leviton and Aldrich, "The Impact of Travels on Scientific Knowledge", pp. 126, 128. Blanford's view also demonstrates the limits of early thinking about mountain formation. It is now understood that the Zagros mountain range is the result of the convergence of tectonic plates.

was the image of a southern landmass that "linked" Africa, South America, India, Australia and New Zealand and another unknown continent, connected to the others, and which "may have occupied the Antarctic area".[37]

The work of the Blanfords, with its early postulation of the sundered connection between landmasses, also set the stage for the convergence of science and myth.

Philip Sclater, a British biogeographer, had undertaken research in the Americas and spent the greater portion of his life as the Secretary of the Zoological Society of London. In 1864 he wrote a paper that attempted to explain the distribution of lemurs. Sclater had a fairly broad perspective of what constituted the species and so found them distributed in areas as diverse and distant as Madagascar, Sri Lanka and south-east Asia. For Sclater the only conclusion that explained how a species could travel from Sri Lanka to south-east Asia was that there had previously existed a continent that united these different regions, which he named Lemuria as homage to Madagascar's round-eyed residents.[38]

Sclater's continent took on a life of its own in science and literature. Just four years later, Ernst Haeckel proposed that Lemuria was the site of the evolution of hominins. It had been just a decade earlier that the fossilised remains of *Homo neanderthalensis* had been discovered in Haeckel's native country. Prominent science fiction writer H.G. Wells adopted this image of a continent that was an ancestral motherland in *Outline of History* as late as 1919. And no less a personage than Friedrich Engels alluded to it in *Labour in the Transition from Ape to Man*. For the Blanfords, Lemuria became the ideal explanation for their observations and, as early as 1873, Henry Blanford wrote a geography textbook for Indian students where he hypothesised a past connection between Africa and India which had been rent asunder by volcanism. This would strike a particular chord in South India.[39]

In Tamil mythology there exists the story of a catastrophic flood. This flood, termed Katalakōl in Tamil, was believed to have submerged the previously extensive region that comprised the Tamil region, confining them to the southernmost portion of the Indian subcontinent. For many Tamil people, Sclater's Lemuria was proof of the previous existence of this lost land and, as recently as 1971, textbooks in Tamil Nadu included this as part of Tamil history. The southernmost point of India is called

[37] Leviton and Aldrich, "The Impact of Travels on Scientific Knowledge", pp. 129, 133.

[38] Nield, *Supercontinent*, pp. 36–38.

[39] Nield, *Supercontinent*, pp. 31, 42–43.

Kanyakumari—"Kumari" being the name of this vanished land. What is particularly interesting and highlights the complex relationship between myth and science is that the very small island that lies offshore is made up of charnockite, a rock created more than half a billion years ago in the earth's interior as a result of the melding together of two landmasses as part of Gondwana. Geologists have called this island "Gondwana Junction". This marks the point at which South Africa, India, Sri Lanka, Madagascar, Australia and East Antarctica were once fused, forming the eastern part of Gondwana.[40]

Plate tectonics has come to be seen as the overarching theoretical framework used to explain the geological processes that shape the surface of the earth and incorporates within it the elements of uniformitarianism. A century ago, however, its antecedent, continental drift, was considered antithetical to the prevailing explanation provided by Hutton and Lyell and, subsequently, marginalised or met with unremitting hostility—even as its adherents continued to increase. Both uniformitarianism and continental drift, however, were the result of different encounters with the earth's geology that both challenged and affirmed existing belief systems.

[40] Nield, *Supercontinent*, pp. 44–47, 258.

CHAPTER 4

Bedrock: Geology and the Shaping of a Nation

The First Visit of the British Association

Following the end of the South African War, the South African Association for the Advancement of Science (SAAAS) was formed in 1903, with a membership that included both the amateur and the professional.[1] It took its inspiration from the British Association and a delegation from that august institution made its first visit to the country in 1905. This would be followed by another visit in 1929.[2] The 1905 visit of the British Association saw a country in the process of being made. Only a few years after the end of the South African War science became the means of mending the rift between English and Afrikaner as well as strengthening ties to Britain. Its apparent neutrality and noble quest for truth made science the perfect vehicle for sustaining this connection. Ravaged by a war that had laid bare the tensions within white society, facing a peace that would marginalise the black majority, educational and cultural institutions were still in their infancy. An exception, however, was in the area of geology. The mineral discoveries of the late nineteenth century had given South Africa impetus in the field and 1905 saw the publication of the first geology

[1] Saul Dubow, *Scientific Racism in Modern South Africa* (Cambridge, Cambridge University Press, 1995), p. 12.

[2] Cf. Saul Dubow, "A commonwealth of science: the British Association in South Africa, 1905 and 1929" in *Science and Society in South Africa,* Saul Dubow (ed) (Manchester, Manchester University Press, 2000).

S. Chetty, *'Africa Forms the Key'*, Cambridge Imperial and Post-Colonial Studies, https://doi.org/10.1007/978-3-030-52711-2_4

textbook in the country. The recognition of South African expertise marked a contrast to the supercilious attitude of the British Association towards a former colony that was yet to achieve dizzying heights in knowledge creation.[3]

The two hundred delegates of the British Association arrived in Cape Town in August 1905 on board the Union Castle ship, *Saxon*. The first gathering was held in the Cape Town City Hall and the inaugural address was made by the Association's president, G.H. Darwin. Upon his conclusion, the Governor of the Cape, Sir Walter Hely-Hutchison, accompanied by loud applause, hoped that the "brotherhood of science" would extend to the relations between Britain and the Dominions. In Darwin's address, science and the spirit of inquiry became linked to a domination over nature but also became the means by which Europeans—and their settler brethren—had been able to exert their supremacy over the rest of the world:

> The dominance of the European race in America, Australasia and South Africa has no doubt arisen from many courses, but amongst these perhaps the chief one is that not only do "we want to know" but also that we are determined to find out.[4]

At the same time, however, progress was not solely the hallmark of the "European race" and Darwin made mention of "an Oriental race" that had recently embraced European thinking and been strengthened by it.[5] This most likely referred to the Japanese who had espoused European modernity and were beginning their ascendance as a modern—and imperial—nation.

An obstacle to progress in South Africa was highlighted by the Reverend W. Cunningham, president of the Economic Science and Statistics section. The settler population had different experiences compared to those of Australia and North America (where indigenous populations had largely been decimated), nor had they the experience of resistance to the implementation of modernity, the "defined conflict between the old and the new" as had occurred in India. What South African settlers did have to contend with in their application of modern European values was the obstacle to industrial and economic growth presented by "natives [who]

[3] Dubow, "A commonwealth of science", pp. 70, 73, 76.

[4] "The British Association in South Africa" in *The Times,* August 16, 1905, Issue 37787, p. 5, London, Times Newspapers Limited, Gale Document Number CS84078864.

[5] "The British Association in South Africa" in *The Times,* August 16, 1905.

have not been schooled to regular habits of work by the discipline of a highly traditional civilization".[6] A further meeting of the section two weeks later reiterated this theme with a contribution by the Deputy Mayor of Johannesburg, H. Pim who addressed the "native problem". To his British audience, Pim pointed out that the "native" was not yet at the stage of development that would allow him "political power". Furthermore, this development could only be achieved through "the interference of the white races".[7]

In an ironic twist, however, the patronising attitude present on the part of Pim was echoed by the British Association in relation to their South African counterparts. In the first section of his opening address, Darwin made clear the relationship of intellectual inequality, "I take it, then, that you have invited us because you want to know what is worth knowing". This unidirectional transmission of knowledge would be evident in the legacy of this visit where, left in the wake of the visit of the British Association, would be "permanent fertilization in the form of stimulated scientific and educational activity".[8] This would be a theme which South Africans would also accentuate in 1929. As the 1905 visit drew to its close, the Lieutenant-Governor of the Transvaal, Sir Arthur Lawley, listed the many interventions that science could make in the economic and social life of South Africa, bringing to bear the optimism of progress in addressing "the prejudices of caste, colour, and race" and promoting development in homage to "the spirit animating the British Association".[9]

Yet, the one area in which South Africa could assert dominance and expertise was that of geology. A common thread running through the Transvaal phase of the visit was mining.

Mining and Empire

Kimberlites are rocks of volcanic origin with their origins deep within the earth's crust. They reach the surface in the form of magma that subsequently cools and hardens. As it continues its journey towards the surface kimberlite magma incorporates other rocks and minerals, including

[6] "The British Association in Cape Town" in *The Times,* August 17, 1905.

[7] "The British Association in the Transvaal" in *The Times,* September 2, 1905, Issue 37802, p. 9, London, Times Newspapers Limited, Gale Document Number CS151449890.

[8] "The British Association in South Africa" in *The Times,* August 16, 1905.

[9] "The British Association in the Transvaal" in *The Times,* September 1, 1905, Issue 37801, p. 8, London, Times Newspapers Limited, Gale Document Number CS134410529.

diamonds. The initial depth at which kimberlite magma comes into being also provides the ideal conditions for the metamorphosis of carbon to diamond. Kimberlite pipes are therefore a source of large diamond deposits.[10] The term "Kimberlite" is derived from Kimberley, the site of extensive diamond diggings in the late nineteenth century. Initially in India and then Brazil, the focus of diamond mining shifted to South Africa in 1867. The almost apocryphal story of the discovery of the first diamond in 1867 is known to most South African schoolchildren. The opinion of an early geologist post-discovery, however, was that it was a fluke or a hoax. Subsequent discoveries—first a trickle then a flood—soon put paid to expert opinion. Within two years of the initial discovery the area around the Vaal River was swamped by prospectors hoping to strike it rich, each setting up competing small claims, first employing their own labour and then that of black workers.[11]

Dr William Atherstone, a medical practitioner, had been instrumental in identifying the first diamond discovered in 1867. At Kimberley he described to future mining magnate, Barney Barnato, the process whereby Kimberlite reaches the surface, carrying with it diamonds. From his description, Barnato surmised that the greater number of diamonds still awaited discovery. By now, much of the stones from the easily accessible yellow earth had been exhausted. Miners now found themselves faced with dauntingly dense blue rock. Convinced that this was where diamonds could still be found, Barnato was able to buy claims from those unable to use new, more expensive methods of excavation.[12] As mining reached greater depths, greater skill and expertise were required and claims were amalgamated by companies able to access the capital necessary to create large monopolies. William Worger demonstrates that, with the growth of the diamond monopolies came a greater control over black labour as these powerful capital interests, personified by men like Cecil John Rhodes, were able to exercise their economic might at the political level and implement controls over black labour, culminating in the housing of workers in

[10] Ludwig-Maximilians-Universitaet Muenchen (LMU) "How diamond-bearing kimberlites reach the surface of Earth: Acidification provides the thrust", 26 January 2012. http://www.sciencedaily.com/releases/2012/01/120123094523.htm, Accessed 9 December 2014.

[11] Wiliam H. Worger, *South Africa's City of Diamonds: Mine Workers and Monopoly Capitalism in Kimberley, 1867–1895* (Craighall, AD Donker, 1987), pp. 2–3, 9–11.

[12] James Leasor, *Rhodes and Barnato: The Premier and the Prancer* (London, Leo Cooper, 1997), p. 95.

closed compounds, arbitrary searches and restriction of movement.[13] Herein lay the origins of the apartheid state. This control would take on even greater proportions on the Witwatersrand.

In 1905, David Draper related the story of John Henry Davis, a "practical mineralogist" of English origin who discovered gold in present-day Krugersdorp in the Transvaal in 1852.[14] Early attempts to locate the extensive gold deposits on the Witwatersrand in the mid-nineteenth century failed due to a lack of sufficient geological knowledge. While gold strikes had been made in the region, geologists tended to focus on checking for traces of the metal in water sources. Geologically, the Witwatersrand is a basin which, three billion years ago, was the site of an ancient sea. Water flowing into this vast body of water carried with it vast quantities of sediment, including gold. Over time, intense heat and pressure transformed these layers of sediment into rock called conglomerates. The mother lode was there, present in these conglomerates, an unusual site of gold deposits, and was only discovered in the late nineteenth century. It would ultimately yield approximately a third of all gold ever mined.[15]

Thus little came of discoveries by early pioneers until 1886 when gold was found in Barberton, precipitating a gold rush and permanently changing the face of South Africa.[16] Like the early days of Kimberley, those attracted to the goldfields were adventurers and dreamers hoping to accumulate an easy fortune. Providence played no small role. E.J. Karrstrom tells the story of a young, unnamed Swedish adventurer who travelled to South Africa in the late nineteenth century. Caught up in the conflict on the eastern frontier, spending a brief period on the diamond fields with little to show for it, his meandering journey eventually took him to Barberton where he began prospecting for gold. Early prospecting was anything but an exact science. It was dependent on basic geological knowledge derived from experience—promising quartz samples were crushed,

[13] For a detailed account cf. Wiliam H. Worger, *South Africa's City of Diamonds: Mine Workers and Monopoly Capitalism in Kimberley, 1867–1895* (Craighall: AD Donker, 1987).

[14] Ethel and James Gray, *A History of the Discovery of the Witwatersrand Goldfields* (Johannesburg, Sholto Douglas and Company, 1940), pp. 2–3.

[15] Jade Davenport, *Digging Deep: A History of Mining in South Africa* (Johannesburg, Jonathan Ball Publishers, 2013), pp. 139–140, 134–136.

[16] Stanley Trapido, "Imperialism, Settler Identities, and Colonial Capitalism: the Hundred Year Origins of the 1899 South African War" in *The Cambridge History of South Africa, Volume 2: 1885–1994,* R. Ross, A.K. Melger and B Nasson (eds) (Cambridge, Cambridge University Press, 2011), p. 82.

washed and inspected for traces of gold.[17] The nature of the gold deposits on the Witwatersrand, however, meant that the ore lay at incredibly deep levels and was of relatively low grade. Extracting the gold was a labour intensive process and dependent on expensive equipment, particularly as mines became increasingly deeper.[18] The days of the individual prospector were numbered.

Those most poised to take advantage of the gold deposits were the mining magnates who had made their fortunes on the diamond fields of Kimberley. Cecil John Rhodes dominated both worlds. Like Barnato, Rhodes—with his partner Charles Rudd—bought up smaller claims on the diamond fields, holding majority shares in the De Beers mine and forming the De Beers Mining Company. In 1887 Rhodes and Rudd went on to establish what was to become Consolidated Gold Fields of South Africa.[19] However, not content with his impressive holdings on the Witwatersrand and fuelled by imperial ambition, Rhodes looked further north to the kingdom of Lobengula of Matabeleland.

The inhabitants of the pre-colonial settlement of Great Zimbabwe had already exploited the gold resources of the area. Gold extraction formed an integral part of their economy, along with agriculture, trade and the rearing of livestock. The extraction of gold was done in the form of mining during the dry seasons while panning for gold took place throughout the year. The trade in gold took on particular importance, forming the basis of the political and economic power of some societies such as the Mutapa state.[20] Nineteenth century accounts by explorers and travellers portrayed Matabeleland as "the Land of Ophir" with abundant gold deposits. Responsible for some of these early descriptions was Carl Mauch whose first viewing of these gold fields had him mesmerised. Mauch was a geologist and was enlisted to gather geographical data in Africa. During his four years in southern Africa, he accumulated rock specimens and searched for signs of mineral wealth.[21] At the same time, Mauch's travels

[17] Cf E.J. Karrstrom, *Eighteen Years in South Africa: A Swedish Gold-Digger's Account of His Adventures in the Land of Gold (1877–1896)* I. Rudner (ed), I and J Rudner (translators) (Cape Town, Africana Publishers, 2013), p. 130.

[18] Davenport. *Digging Deep*, p. 295.

[19] Leasor, *Rhodes and Barnato*, pp. 117–118, 143.

[20] Gerald Chikozho Mazarire, "Reflections on Pre-Colonial Zimbabwe, *c.* 850–1880s" in *Becoming Zimbabwe: A History from the Pre-colonial Period to 2008,* Brian Raftopoulos and Alois Mlambo (eds) (Harare, Weaver Press, 2009), p. 37.

[21] Davenport, *Digging Deep*, p. 79.

were clandestine as Mzilikazi, the chief of the Ndebele,[22] prohibited mining. Mzilikazi's death and the succession of a more amenable leader in the form of Lobengula, however, paved the way for a steady stream of prospectors. Both small concerns such as the Durban Gold Mining Company as well as the more substantial London and Limpopo Mining Company tried in vain to locate gold on the scale envisaged by Mauch. Despite the lack of success, the rumours of fabulous gold deposits persisted.[23] Inspired by the belief that Matabeleland was a "Second Rand", Rhodes' British South Africa Company was granted a concession to pursue mineral resources in the region. With sweeping military and political power, the concession paved the way for the ultimate colonisation of the land that would be named after its imperial architect, Rhodesia.[24]

Like Zimbabwe, Namibia—then termed South West Africa—was also the target of mineral speculation. Most of the mineral rights of the territory were held by the Deutsche Colonial Gesellschaft fur Sudwest Afrika. Other concession companies were soon involved and granted the right to extract minerals by the German government. Individuals too were granted concessions by chiefs to mine areas under chiefly control. In 1899, De Beers became involved in the area, a move that was prescient when diamonds were discovered in South West Africa in 1908. The diamond deposits were extensive and by the outbreak of the First World War, South West Africa was producing one-fifth of the world's diamonds. Also mined were tin and copper and, later in the century, uranium deposits were discovered as well. When the First World War broke out, South African forces invaded South West Africa—ostensibly for reasons of security—and assumed control over the territory in July 1915. By 1919, South African control was formalised when South West Africa became a South African mandate.[25] A step had been taken towards fulfilling Jan Smuts's vision of South African leadership on the continent.

[22] Matabele in the early nomenclature.

[23] John S. Galbraith, *Crown and Charter: The Early Years of the British South Africa Company* (Berkeley and Los Angeles, University of California Press, 1974), pp. 31–32.

[24] Sabelo J. Ndlovu-Gatsheni, "Mapping Cultural and Colonial Encounters, 1880s–1930s" in *Becoming Zimbabwe: A History from the Pre-colonial Period to 2008,* Brian Raftopoulos and Alois Mlambo (eds) (Harare, Weaver Press, 2009), p. 46.

[25] William U. Crowell, "The Evolution of South African Control over South West Africa (Namibia)" (St John's University, Unpublished PhD Dissertation, 1975), pp. 33–35, 66–67, 74, 95.

In an industry that had produced the likes of Rhodes and Barnato, Ernest Oppenheimer would stand out as the foremost of South Africa's mining magnates. Of German Jewish extraction, Oppenheimer immigrated to England as a teenager and worked as a diamond sorter in London. He was 22 when he reached Kimberley where he quickly acquired a reputation due to his expert knowledge of diamonds. The firm for which he worked—Dunkelsbuhler—had close ties to De Beers. The First World War saw him go back to England in 1915 but he returned a year later—to the Witwatersrand. After raising money from influential international financiers, Oppenheimer established the Anglo American Corporation of South Africa. He turned his attention to South West Africa, using contacts he had made with prominent South Africans—Smuts being a prime example—to obtain information regarding the prospective sale of German mines at the conclusion of the First World War. As such, Anglo American was able to purchase 11 German mining companies in South West Africa, creating Consolidated Diamond Mines of South West Africa Limited which had a virtual monopoly over diamond production. Oppenheimer established himself as a giant in the gold and diamond mining sectors, ultimately using the resources of Anglo American to purchase shares in De Beers until he was eventually named a director of his one-time rival, becoming chairman in 1929.[26]

The policies implemented by the South African government in South West Africa regarding mining and the control of black labour—as well as land dispossession—had their nascence in South Africa. In the aftermath of the South African War, the mining industry was perceived to be integral in rebuilding the shattered South African economy. Alfred Milner, the South African High Commissioner, who was also named the Governor of the Transvaal, viewed black labour as key to reconstruction. It would also have the additional benefit of a "civilising" effect. The creation of a South African union or federation had been a British vision since the nineteenth century. Its advantages in terms of political administration, the creation of infrastructure and economic exploitation were self-evident. The South African War had provided the ideal conditions for the realisation of British—and mining—ambitions. The creation of the Union of South Africa in 1910 retained the political disenfranchisement that had applied to black South Africans in the former Boer republics and Natal with the limited property qualified franchise still in existence at the Cape. It

[26] Davenport, *Digging Deep*, pp. 249–253, 260–261, 263, 265, 270, 276.

shattered the aspirations of the emerging black middle class and would have the long-term effect of confining black South Africans to the migrant workers of a capital-intensive mining industry that required an acquiescent and cheap labour force, the control of which drove the discriminatory legislation of segregation in the 1920s and, ultimately, apartheid in 1948.[27]

Mining had therefore been integral to a particular concept of South African nationhood and its position within the British Empire. It was the site of contestation that would define South African history. It was also the means by which South Africa could assert expertise—the challenges of gold extraction in particular fostered engineering and chemical innovation. and it was therefore in the realm of geology where South Africa was not merely the passive recipient of British proficiency. This, however, was not necessarily recognised by the British Association. The speech made by the president of the Geological section, Professor H.A. Miers, was focused on the deposition of ores with a particular emphasis on crystallography. It was, however, theoretically based with little consideration of practical concerns related to extraction. This allowed him to affirm Britain's intellectual leadership: "we might hope that the visit of the British Association would be of some help to her younger sister in the task of diffusing a taste and an interest for the pure truths of science".[28] It is within this context that the 27-year-old Du Toit made his presentation—which was summarised in detail in the British press.

Du Toit's paper focused on a description of the Stormberg Formation, which he had spent the previous two years mapping. As Du Toit pointed out, this formation was extensive, apparent in Natal, the Orange River Colony and Basutoland—as well as the Cape—and was a record of an increasingly arid climate. He described its composition—four types of strata including volcanic rock and sandstone. Of particular interest were the thin coal beds—the only deposits in the colony where extraction was economically viable. In a precursor to the work that he would do later, Du Toit paid particular attention to the fossils that characterised this formation—both plant fossils and those of dinosaurs such as *Massospondylus*.[29] It was also here that Du Toit began tentatively to highlight the similarities

[27] Shula Marks, "War and Union, 1899–1910", pp. 167–168, 190–194.

[28] "The British Association in at Victoria Falls: Opening of the Bridge" in *The Times,* September 13, 1905, Issue 37,811, p. 9, London, Times Newspapers Limited, Gale Document Number CS151187757.

[29] "The British Association in South Africa" in *The Times,* August 18, 1905.

between the Karoo Supergroup and those in India and in New South Wales in Australia.[30]

While the extensive geological description may have appeared somewhat dry to someone outside the field, where Du Toit excelled was in his evocation of past worlds. There was the vast inland sea sited in the present-day Karoo, the sediment of which comprised the Karoo system. There were the extensive volcanic eruptions, leading to layers of lava covering the sandstone deposits. According to Du Toit, the remnants of more than a hundred volcanoes had been detected, a few of which were massive. Evident too were changing mountainous landscapes—the rise of the region in the interior and the subsidence in the south that would eventually vanish beneath the waves of the Indian Ocean.[31] It was the work of the imagination—to reconstruct the alien worlds of the past from the fragments available in the present and it was this imaginative impulse that would later compel him to create an entire continent.

Du Toit's Scientific Method

Du Toit's ability to recreate a lost world was based on a particular approach to science. In a sense, it is ironic that his work paralleled so closely that of Alfred Wegener who is most closely associated with continental drift. Geologists had considered Wegener, an outsider and, even worse, an "armchair geologist"—despite all evidence to the contrary. Gevers adopts a patronising tone towards those scientists content to arrive at deductions based on the work of "other more enterprising and adventurous adherents".[32] It was clear that Du Toit was the very antithesis of the "armchair geologist". A geography textbook written by him in 1912 advises his young readers that:

> Too much stress cannot be laid on the importance of actual observational work out of doors to supplement the teaching of textbooks, while, wherever

[30] T.W. Gevers. "The Life and Work of Dr. Alex. L. Du Toit"—Alex L. Du Toit Memorial Lectures No. 1. In *The Geological Society of South Africa: Annexure to Volume LII* (Johannesburg, The Geological Society of South Africa, The South African Association for the Advancement of Science, The South African Geological Society, 1949), p. 47.

[31] "The British Association in South Africa" in *The Times,* August 18, 1905.

[32] Gevers, "The Life and Work of Dr. Alex. L. Du Toit", p. 36.

possible, photographs, sketches and plans of instructive features and areas should be taken.[33]

Du Toit's work was founded on "facts", the myriad observations he made in the field which were subsequently preserved in his many notebooks. Nor, too, were these facts solely his own. He had a tendency to accumulate items of interest from other researchers, snippets of information that would be preserved in the hopes that they may form some missing piece of the puzzle. It was inductive reasoning at its purest—the assemblage of raw data in sufficient quantity that it may eventually allow a "truth" to be discovered:[34] "I have ever set a value on those little scraps of knowledge which my fellow geologists have now and then disgorged, since their potential worth could seldom be straightaway assessed".[35]

In this quest for geological knowledge and his prioritisation of fieldwork, Du Toit made use of increasingly sophisticated transportation technologies ranging from a donkey wagon to a push-bike, trains and, ultimately aeroplanes.[36] It is wonderfully symmetrical to note the ways in which, just as Du Toit was able to travel from the local to the global so, too, did the breadth of his vision increase until it encompassed the entire planet. He was not unique in this. The vision afforded from the air—and in the 1960s, from space—allowed for a different perspective. Alfred Wegener, too, had had a passion for flight, using hot air balloons in the course of his meteorological research.[37]

It was this ability to construct a whole, what Gevers contradictorily terms "intuition", that underlay Du Toit's work in the field. From very early on in his fieldwork, continental drift served as the interpretative framework for the "facts". Being meticulous in the field then complemented an instinctive "intuition" that was compounded by an openness to new ideas.[38]

[33] Alex L. Du Toit, *Physical Geography for South African Schools* (Cambridge, Cambridge University Press, 1926), p. v.

[34] T.W. Gevers, "The Life and Work of Dr. Alex. L. Du Toit", p. 37.

[35] Alex L. Du Toit, "Some Reminiscences" in *Transactions of the Geological Society of South* Africa, Jubilee Volume, 1946 reprinted in T.W. Gevers, "The Life and Work of Dr. Alex. L. Du Toit", p. 38.

[36] T.W. Gevers, "The Life and Work of Dr. Alex. L. Du Toit", p. 38.

[37] Robert Muir Wood, *The Dark Side of the Earth* (London, George Allen and Unwin Publishers Ltd., 1985), pp. 10–11.

[38] T.W. Gevers, "The Life and Work of Dr. Alex. L. Du Toit", pp. 38–39.

If one moves beyond the individual attributes of personality and intellect, Du Toit was also a South African, a member of a fledgling nation, rent by division and inequality, that was attempting to assert an identity and a sense of space, both geographically and intellectually, and both would come together in his work.

Five years after the visit of the British Association, the Union of South Africa came into being on 31 May 1910, consisting largely of the amalgamation of two former Boer republics—the Transvaal and the Orange Free State in the north—and the two British colonies of Natal and the Cape. This union was reflected two years later when the Cape and Transvaal Geological Commissions were joined to form the Geological Survey of the Union of South Africa, with A.W. Rogers assuming the directorship of the new association. The merger did little to affect the fieldwork activities of Du Toit who still remained the youngest geologist in the Geological Survey.[39]

As well as his study of the Karoo succession, Du Toit's fieldwork also encompassed the location and availability of underground water deposits—which was of particular importance for South Africa, a largely arid country plagued by periodic droughts that affected the availability of surface water and a new state that also had to contend with burgeoning development in industry and commercial agriculture.

By 1906, Du Toit had written a paper entitled "Underground Water in South-east Bechuanaland" and in 1907 he provided guidance on where to place boreholes in the more arid northern regions of the country so as to access underground water. In 1913, he presented a paper to the South African Society of Civil Engineers, "The Geology of Underground Water Supply with Special Reference to South Africa". His focus on the Karoo succession meant that he also studied the way in which the dolerite within this section intruded, while molten, into the surrounding rock forming dykes and sills that subsequently influenced the motion of groundwater.[40] These papers helped establish him as an authority on underground water supply or hydrogeology—this, in addition, to his acknowledged expertise in palaeobotany, petrology and mineralogy.

At the beginning of 1914, then, Du Toit was able to combine his various geological interests on a trip to Australia. Part of the motivation for his visit was the opportunity to analyse rock strata that was of the same age as

[39] Gevers, "The Life and Work of Dr. Alex. L. Du Toit", p. 14.
[40] Gevers, "The Life and Work of Dr. Alex. L. Du Toit", pp. 13–14, 19, 69.

the Karoo System. This may suggest that he had already begun to consider comparisons between South African stratigraphy and those of other Gondwana remnants.[41] Du Toit's notebook recording his Australian visit is an amalgam of detailed descriptions of rock formations, type and stratigraphy accompanied by rough sketches. Their descriptive nature provides the model for his later published work and, at the outset, there is little evidence of conclusions drawn. A tantalising hint, however, is provided in his account of Hallet Cove in South Australia where "here is no doubt that … the ice came from the SE from a great landmass in which granite predominates".[42] A possible very early reference to Gondwana and indicative of the theoretical path that Du Toit was already beginning to take, Du Toit's preliminary assessment was borne out in subsequent understanding of the area's geology. A significant geological site, Hallet Cove contains the remnants of three periods of glaciation. The most recent occurred during the Permian era approximately 280 million years ago. This glacier apparently covered most of Gondwana, including Australia which subsequently began its northerly movement away from Antarctica to which it had once been joined. The evidence of striations made by the moving ice sheet can be seen in Black Cliff that forms part of the Hallet Cove Conservation Park.[43]

Du Toit's notebook also highlighted his role as a hydrogeologist with reference to water pressure and salinity[44] and he spent part of his time in Australia investigating the Great Artesian Basin.[45] This Basin is a truly immense reservoir of groundwater that underlies a substantial portion of the Australian continent. It holds almost 75 million megalitres of water, some of which may be two million years old. Running beneath some of the dry and desert regions of Australia, the Basin is able to supply much needed water to support both the civilian population as well as

[41] Gevers, "The Life and Work of Dr. Alex. L. Du Toit", p. 19.

[42] UCT-JL: Alex L. Du Toit Papers—BC 722: L: Alex Du Toit material transferred from Geological Sciences, Oct 2005—Field Notebooks: 1914—Australian Notes.

[43] Mark Willoughby "Overview of Hallet Cove Geological History" https://www.mindat.org/article.php/957/Overview+of+Hallett+Cove+Geological+History, Accessed 18 October 2017.

[44] UCT-JL: Alex L. Du Toit Papers—BC 722: Field Notebooks: 1914—Australian Notes.

[45] Gevers, "The Life and Work of Dr. Alex. L. Du Toit", p. 19.

industry[46]—something in which Du Toit was particularly interested due to the growing industrialisation in South Africa.

Du Toit was intensely conscious of the variable nature of South Africa's water supply as a result of erratic rainfall, believing that South Africa was in a particularly poor position in relation to water supply than other countries. In addition to inadequate rainfall, and the limitations of water provided by rivers, there were human agricultural activities that led to soil erosion that had a detrimental effect on groundwater. An inadequate water supply, moreover, had the potential to hinder South Africa's industrial and agricultural development, so essential for a young, modernising nation. Du Toit also believed that science could be the panacea for the country's water woes and would later prove to be instrumental in initiating the Water Branch of the Union Geological Survey, with an emphasis on the use of windmills and pumps, dams to store water and the provision of knowledge to reduce soil erosion and ensure groundwater quality. For Du Toit, if necessary, ordinary South Africans should be "compelled" to make use of scientific knowledge in a manner that would minimise damage to the environment and it was science that would be key to South Africa's progress: "it is to science that the country must turn for its rehabilitation … the scientist should take a more active part in guiding the affairs of the nation".[47] This optimistic view of science was little diminished when it became associated with war.

Prior to the outbreak of the First World War, Du Toit demonstrated his formidable knowledge of South Africa's water supply in an exhaustive survey. Combining geology and climate, the paper was published in the journal *Civil Engineering* and contained information relevant to the planned drilling of boreholes and other means of accessing groundwater. Du Toit begins, however, by contextualising South Africa's water supply within South African exceptionalism—the country's unique geology, climate and topography implying that it is distinct from conventional European scientific understanding:

> South Africa has proved so peculiar in matters scientific as judged from European standpoints, that it is no surprise to find the local problems of

[46] "The Basin" Great Artesian Basin Coordinating Committee, http://www.gabcc.gov.au/basin, Accessed 18 January 2017.

[47] Gevers, "The Life and Work of Dr. Alex. L. Du Toit", pp. 71–74.

> hydrogeology to be in many respects different from those dealt with in current literature.[48]

It is an assertion of South Africa's unique nature, the country's individuality that cannot be pigeonholed by European scientific thought. At the same time, Du Toit is less than enamoured by the lack of attention allocated to this important issue by the state—a disregard that is all the more glaring due to the country's perennial problems with water supply. Unsurprisingly, with his expertise in geology, Du Toit links the groundwater supply to the characteristic rocks, citing the overall preponderance of non-porous rocks for the poor groundwater availability. Rocks such as sandstone have a higher degree of porosity, allowing the water to seep through, but these rocks are largely found in the Cretaceous System and are not dominant in South African geology, with the exception of very few regions. Compounding the lack of groundwater is the limited run-off with a rather small percentage of South Africa's rainfall feeding the country's rivers. Most of the rainfall is lost to evaporation.[49] It is therefore a convergence of geological and climatic conditions that have adversely affected the water available in the country—a resource that is necessary for development, a central concern for Du Toit. In a subsequent response to the discussion initiated by his paper, Du Toit encouraged further collaboration between geologists and engineers with the aim of "accomplishing schemes that may advance the welfare of the Union of South Africa".[50]

His paper advises engineers on suitable sites for the drilling of boreholes, highlighting the various rock formations around the country and their effect on water supply and quality. The dolomite characteristic of the Transvaal, for instance, has resulted in the permeation of water below ground with a lack of surface water. Moreover, he challenges the prevailing view of dolomite which had long discouraged engineers as a potential source of water. While this was the case with newer dolomite formation, older rocks such as those found in Natal allowed water to seep through cracks and fissures brought on by weathering. Water quality was also affected by environmental and geological conditions such as the greater

[48] Alex L. Du Toit "The Geology of Underground Water Supply with Special Reference to South Africa" in *Civil Engineering*, Vol 1, pp. 8–31, 1913. http://www.sabinet.co.za Doc AJA10212019_185,214, Accessed 13 October 2016, p. 8.

[49] Du Toit, "The Geology of Underground Water Supply", pp. 8, 10, 11–12.

[50] Du Toit, "Reply on Underground Water Supply", Fourth Meeting, 11th June 1913, Sabinet Doc AJA10212019–185214, Accessed 13 October 2016, p. 84.

salinity found in areas with less rainfall as well as "dips" in the rock strata. Du Toit advised the drilling of boreholes at shallow depths where water may be found with the drilling of deep boreholes judged an expensive enterprise that bore little fruit.[51]

Pre-empting his work during the First World War, Du Toit extended his knowledge to German-controlled South West Africa where he argued for good groundwater supplies due to the Karoo sandstone which abounds in this area. He was optimistic about the Kalahari as a whole as a site with the potential "for stock farming and settlement".[52] Du Toit's work in hydrogeology can be contextualised by growing attempts by the state to address the inadequate water supply for agriculture, the use of scientific farming methods and the contribution of drought to poverty and, in particular, white poverty. These attempts were evident in the work of another Du Toit, Heinrich S. Du Toit, who would go on to chair the Drought Investigation Commission in 1920. In 1914, after Alex Du Toit presented his work on underground water supplies, the periodic droughts and its adverse effects were a significant government concern, leading to the establishment of the Senate Select Committee.[53]

The year 1914 proved to be a significant year for South Africa (and the world), marking the outbreak of the First World War. It was a conflict that laid bare the tensions within white society, only four years after unification. In 1912 in his capacity as Minister of Defence, Jan Smuts, had introduced the South African Defence Bill that attempted to forge a united Union Defence Force out of the two different systems—the British regimental system and the Boer commando system. Passed as the South Africa Defence Act, the legislation aroused both controversy and opposition not least of which was its attempts to create a modern military system that alienated Boer veterans accustomed to a more personal system of patronage with different perceptions of discipline. Upon the outbreak of war, South Africa was to confront German forces in the German colony of South West Africa yet a number of senior Afrikaner military leaders resigned from their positions in protest of South Africa's

[51] Du Toit, "The Geology of Underground Water Supply", pp. 14, 16, 22, 24.

[52] Du Toit, "The Geology of Underground Water Supply", p. 28.

[53] William Beinart, *The Rise of Conservation in South Africa: Settlers, Livestock, and the Environment 1770–1950* (Oxford, Oxford University Press, 2003), Chapter 7.

participation in the war subsequently leading a rebellion against the Union government.[54]

Du Toit, however, had little compunction in serving in his capacity as a hydrogeologist. An important requirement for South African troops in South West Africa was the provision of water in a semi-arid region. With the outbreak of war, Du Toit was transferred to the Defence Force as a captain, with the responsibility of seeking out suitable borehole sites in the southern Kalahari Desert. Du Toit also initiated a system of record keeping—listing the statistics related to the type of rock and depths at which groundwater was located as well as both the amount of water and its purity. This information was later expanded upon but remained at the core of further groundwater analysis. Du Toit's war service lasted as long as the South West African campaign and he was discharged in 1915.[55]

His time spent on the campaign, however, also allowed him to examine the strata through which the boreholes were drilled. Geologists tend to be restricted to outcrops—the parts of the rock that are exposed at the surface. Boreholes provided a unique opportunity and Du Toit made extensive notes of the stratigraphy, publishing at least two papers over the next year relating to this aspect of the Karoo System. Du Toit discovered that the Ecca Group which made up part of the Karoo Supergroup and ordinarily consisted of blue shale, derived from sedimentary layers of mud, took on a different form along the Molopo River in the Kalahari. Here, the Ecca Group consisted of alternate layers of sandstone and shale and the shale tended to be purple, green and red rather than blue. Du Toit called this the "Kalahari Phase" or "Red Ecca" which could be distinguished from the more conventional "Cape Phase" blue Ecca. In his later work on South America, he discovered similar "Red Ecca" in Brazil which would be of particular importance in demonstrating continental drift.[56]

Yet another piece of early evidence came from petrology which is the field of geology related specifically to rocks in terms of their origin and chemical make-up. While Du Toit was not by training a petrologist, he demonstrated in this subfield—as in so many others—a more than passing familiarity with the subject matter. He wrote on intrusive granites, ancient igneous rock derived from volcanic eruptions that formed part of the

[54] Ian van der Waag, *A Military History of Modern South Africa* (Johannesburg and Cape Town, Jonathan Ball Publishers, 2015), pp. 74–76, 95–100.

[55] Gevers, "The Life and Work of Dr. Alex. L. Du Toit", pp. 19, 69.

[56] Gevers, "The Life and Work of Dr. Alex. L. Du Toit", pp. 19–20.

Karoo System, metamorphic rocks and charnockite. Discovered and named in South India, Du Toit was the first to discover this form of granite in South Africa.[57] Charnockite is a largely metamorphic rock of granitic origin and is formed at high pressure and fairly high temperatures[58]—like those found where continents were once joined together.

The extensive fieldwork characteristic of Du Toit's early career and his ability to immerse himself in different areas of geology can be seen as the assembly of a great number of pieces of a jigsaw puzzle. With time, he would begin to put these pieces together. As Gevers mentions repeatedly in his biography, Du Toit possessed a "scientific outlook [that] was holistic".[59] He had a need to create order and structure, to assemble a whole from what appeared to be disparate parts.

By 1920, Du Toit had entered a new phase in his career which allowed him greater time to reflect on the years of fieldwork. His extensive knowledge of hydrogeology meant that he had been serving as an advisor to the Union Irrigation Department. The relationship was formalised in 1920 and, no longer burdened by the demands of constant fieldwork, Du Toit was able to assemble the data he had gathered over the years, publishing prolifically. This culminated in the *Geology of South Africa* in 1926, a seminal volume that was the product of more than two decades of observations and deductions. It was, however, during the years leading up to this that Du Toit began to present publicly his views on continental drift.[60]

[57] Gevers, "The Life and Work of Dr. Alex. L. Du Toit", pp. 62–63.
[58] https://global.britannica.com/science/charnockite, Accessed 23 January 2017.
[59] Gevers, "The Life and Work of Dr. Alex. L. Du Toit", p. 21.
[60] Gevers, "The Life and Work of Dr. Alex. L. Du Toit", pp. 25–26.

CHAPTER 5

On the Shoulders of Giants: Early Drift Theorists

When Alfred Russel Wallace, independently of Charles Darwin, arrived at the notion of evolution through the process of natural selection, it was clearly a theory whose time had come.[1] To an extent, the same could be applied to continental drift. While Alfred Wegener is the figure most associated with continental drift, he built upon the work of geologists in the nineteenth century, the most prominent of whom was Eduard Suess. Like-minded contemporaries of Wegener were American geologist Frank Bursley Taylor and, of course, Alex Du Toit, all of whom claimed to have arrived at continental drift as an explanation for geographical and palaeontological congruence across continents. Despite the controversy it aroused, only leading to its acceptance decades later, this, too, was clearly an idea whose time had come.

Eduard Suess

Eduard Suess, born in London in 1831, was the son of a wool merchant who moved back to Prague in 1834. His family placed a premium on education and a private tutor was engaged prior to Suess entering grammar

The phrase used in the title is accredited to Sir Isaac Newton

[1] This paraphrases the quote accredited to Victor Hugo, "An idea whose time has come".

S. Chetty, *'Africa Forms the Key'*, Cambridge Imperial and Post-Colonial Studies, https://doi.org/10.1007/978-3-030-52711-2_5

school. His family subsequently moved to Vienna and Suess joined the Polytechnikum (Technical University) when he was 16.[2] He was still in his teens when he developed an interest in geology. His route to geology was via palaeontology and Suess was initially captivated by the fossilised specimens on display at the present-day Národní Muzeum in Prague and their evocation of bygone worlds. This led him to employment at the Imperial Natural History Cabinet in Vienna. The rocks in which the fossils were embedded indicated different environments captured in the strata. For Suess, this suggested global changes and he envisaged a far more ambitious process of transformation than Lyell's uniformitarianism or Darwin's regional vision of evolutionary competition and adaptation. The study of mountains would provide Suess with his answer.[3]

Suess' rise in the natural sciences was impressive. When he was 26, the Austrian Emperor Franz Joseph I gave him an associate professorship in palaeontology at the University of Vienna. A decade later, he was appointed Professor of Geology.[4] Suess was also part of the Imperial Geological Survey,[5] a by-product of the Industrial Revolution sweeping the European continent where the location and extraction of mineral resources were matters of economic importance.[6] Here, Suess was able to capitalise on his love for mountaineering when he worked alongside Franz von Hauer, an Austrian geologist, mapping part of the Alps. Suess' study of the Alps as well as his observations made of other European mountain ranges such as the Carpathians caused Suess to take issue with the contemporary view of orogeny. Mountain-building was considered to be the result of a catastrophic global event rather than as a result of a gradual process. Suess combined the two extremes—mountain-building could be a local process but, at the same time, geology was influenced by global changes which allowed for the deposition of similar strata containing the same fossil

[2] Thomas Hofmann, "Milestones of a Life beyond the Geosciences" in *The Face of the Earth: the Legacy of Eduard Suess*, T. Hofmann, G. Bloschl, L. Lammerhuber, W.E. Piller, A.M. Celal Sengor, (eds) (Munich, European Geosciences Union, 2014) pp. 42–43.

[3] A.M. Celal Sengor "Eduard Suess and the Origin of Modern Geology" in *The Face of the Earth: the Legacy of Eduard Suess*. T. Hofmann, G. Bloschl, L. Lammerhuber, W.E. Piller, A.M. Celal Sengor, (eds) (Munich, European Geosciences Union, 2014) pp. 14–15.

[4] Hofmann, "Milestones of a Life beyond the Geosciences, p. 42.

[5] Celal Sengor, "Eduard Suess and the Origin of Modern Geology", pp. 16–17.

[6] Ted Nield, *Supercontinent: Ten Billion Years in the Life of Our Planet* (London, Granta Books, 2008) p. 63.

material across continents.[7] Suess termed these global events "transgressions", marked by a rise in sea level and "regressions", a decline. Fossil evidence was of either deep-water or coastal marine fauna which was an important indicator of sea level and its associated environment and climate. For Suess, this rising and falling sea level could be explained by subsidence. The subsidence of the sea floor was the means by which "regressions" occurred and the changes could be read in the fossil record. The environmental changes caused by these global-scale events were key as they promoted particular evolutionary changes that could be determined over large areas. In this way, Suess combined palaeontology, stratigraphy and the understanding of evolution in the late nineteenth century.[8]

His perusal of the fossil record and hypothesis of large-scale evolutionary change as a result of "transgressions" and "regressions" allowed Suess to postulate the existence of large landmasses that he named Angaraland and Gondwanaland, both of which incorporated existing continents and could be defined by their unique flora:[9]

> If we attempt to make similar comparisons with regard to the united mass of Asia, Africa, and Europe, it at once becomes evident that heterogeneous regions—the limits of which do not coincide with the recognized boundaries of these subdivisions—have here been welded together to form one great continent. The first region comprises the southern and a great deal of the more central part of Africa, then Madagascar and the Indian peninsula. The lofty table-lands of this region have never, so far as we know, been covered by sea since primitive times, or at the end of the Carboniferous period; it is only at the foot of the table-lands that marine sediments have been deposited, which followed the encroachment of the Indian Ocean, as this was formed by subsidence within the tabular mass. We call this mass Gondwana-Land, after the ancient Gondwana flora which is common to all its parts.[10]

[7] Celal Sengor, "Eduard Suess and the Origin of Modern Geology", pp. 16–17.

[8] Werner E. Piller, "From Palaeontology and Stratigraphy to Earth System Science" in *The Face of the Earth: the Legacy of Eduard Suess*, T. Hofmann, G. Bloschl, L. Lammerhuber, W.E. Piller, A.M. Celal Sengor, (eds) (Munich, European Geosciences Union, 2014), pp. 21–22.

[9] Piller, "From Palaeontology and Stratigraphy to Earth System Science", p. 22.

[10] Eduard Suess, *The Face of the Earth Vol I* (Oxford, Clarendon Press, 1904), pp. 595–596 reprinted in *The Face of the Earth: the Legacy of Eduard Suess*. T. Hofmann, G. Bloschl, L. Lammerhuber, W.E. Piller, A.M. Celal Sengor, (eds) (Munich, European Geosciences Union, 2014), p. 69.

In addition to the naming of Gondwana, Suess also hypothesised the existence of a gigantic ocean that surrounded the continent which he called Tethys. He believed that the Mediterranean was a relic of this ancient ocean. The significance of his hypothesis was the temporality of oceans—they could be created and destroyed over time and were not permanent features of the Earth and the changing nature of the ocean beds through the process of subsidence was the means by which the "face of the Earth" was altered.[11]

It is, however, as an early proponent of continental movement that Suess acquired particular renown. The mountainous regions in the south of Italy were a fertile ground for teaching aspiring geologists and it was while studying the Apennines that—in conjunction with his existing knowledge of other European mountainous areas—Suess concluded that mountains were not as evenly proportioned as had been supposed. European mountains seemed to display a particularly northward trend as if a force were being exerted on them in that direction. For Suess this suggested that the earth's crust—or at least that portion that comprised Europe—was not stationary.[12] He presented his findings before the Imperial Academy of Sciences in Austria in 1873.[13]

For Suess, mountain-building was a long process that took place incrementally rather than a quick, catastrophic event and their creation appeared to be halted by ocean basins—whether it was the mountains of Europe, the Appalachians of North America, the Cape Mountains of South Africa or the Sierra de la Ventana in South America, they terminated abruptly at the ocean.[14] It is no surprise that the mountains of the southern hemisphere would be integral to the work of Wegener, Keidel and Du Toit in their reconstruction of Gondwana. Suess also considered the stratigraphic similarities across continents that supported the previous connection of these continents.[15]

[11] Piller, "From Palaeontology and Stratigraphy to Earth System Science", p. 23.

[12] A.M. Celal Sengor, "Suess and the Dynamics of the Planet Earth" in *The Face of the Earth: the Legacy of Eduard Suess*, T. Hofmann, G. Bloschl, L. Lammerhuber, W.E. Piller, A.M. Celal Sengor, (eds) (Munich, European Geosciences Union, 2014), p. 29.

[13] Eduard Suess "über den Aufbau der mitteleuropäischen Hochebirge" ("On the Structure of the Middle-European High Mountains"), 17 July 1873 in A.M. Celal Sengor, "Suess and the Dynamics of the Planet Earth", p. 29.

[14] Celal Sengor "Suess and the Dynamics of the Planet Earth", p. 30.

[15] Franz Neubauer, "Gondwana-land Goes Europe" in *Austrian Journal of Earth Sciences*, Vol 107/1, 2014, p. 147.

The movement of continents, however, was one of the most controversial explanations for the common fossil and rock evidence found on the southern continents. As Suess was proposing the connections between continents, Melchior Neumayr—Suess's son-in-law[16]—used the same evidence as support for the existence of "land bridges". The oceans signified continents that had disappeared beneath the depths but which had once connected the various landmasses.[17] Suess, too, privileged subsidence as an explanation for similarities across continents. The earth had been extremely hot at its inception and, as it cooled over time, it shrunk which led to the submergence of land. During this process of shrinking, pressure was put on the Earth's crust which culminated in massive subsidence, dooming large landmasses to the watery depths while allowing for others to remain elevated—these are the landmasses we see today. Simultaneously—and uneasily—Suess had to allow for movement of land laterally that was evident in mountain ranges. This, however, was not a primary movement but a repercussion of vertical movement.[18] For many earth scientists in the late nineteenth and early twentieth centuries, the "land bridge" explanation was preferred and the notion of moving continents was absurd.

This privileging of drowned lands as an hypothesis may be traced even further into the distant past where the boundaries between myth, memory, history and the collective unconscious are blurred. Approximately ten millennia ago, as the glaciers that covered much of the Earth's surface retreated to the poles, the ending of the Ice Age was marked by extensive flooding, inundating coastal and low-lying regions as seas levels rose. Many of the world's religions made reference to a "great flood" from the very early Babylonian myths to the monotheistic regions originating in the Near and Middle East. Legends of mythical lands that had vanished between the waves such as that of Atlantis—to which Du Toit would refer—originated in ancient Greece but continue in some form into the present—albeit often relegated to the realms of pseudoscience.[19] It is therefore ironic that the birth of modern geology and its emphasis on uniformitarianism as a means of distinguishing it from the Biblically tainted catastrophism of early geologists, nevertheless retained elements of

[16] Hofmann, "Milestones of a Life beyond the Geosciences, p. 43.
[17] Newbauer, "Gondwana-land Goes Europe", p. 147.
[18] Nield, *Supercontinent*, pp. 69, 80.
[19] Nield, *Supercontinent*, p. 59.

mythology in "land bridges" that could be traced to the Judaeo-Christian tradition.

Suess's research eventually culminated in the writing of *Die Enstehung der Alpen* or "The Origin of the Alps" which was published in 1875 and the seminal *Das Antlitz der Erde* ("The Face of the Earth"), a mammoth undertaking of four volumes published between 1883 and 1909.[20] His detractors notwithstanding, Suess' work would leave a lasting legacy on the formulation of drift theory in the early twentieth century.

Frank Bursley Taylor

Volume 21 of the *Bulletin of the Geological Society of America* carried a lengthy 47-page article by American geologist, Frank Bursley Taylor. This was based on an earlier abstract submitted 18 months previously at the end of December 1908. While Alfred Wegener is most associated with continental drift, this article by Taylor published on 3 June, 1910, pre-empts Wegener's work by two years.[21] In *A Geological Comparison of South America with South Africa* Alex Du Toit makes reference to both Taylor and Wegener as the early proponents of continental drift, a theory that he had also arrived at independently.[22]

The main part of Taylor's geology education came from his experience in the field. Before completing his studies in geology and astronomy at Harvard University, Taylor left university and began acquiring his expertise as a field geologist in the Great Lakes area. Working on the history of glaciation here, Taylor came to understand that the lakes were created as a result of "ice dams" due to glaciation. He further posited that the sporadic retreat and advance of glaciers over the course of time was due to the landscape of the Great Lakes region—for Taylor it was the land that influenced glacial movement rather than glaciers which had shaped the landscape. His findings were published as *The Pleistocene of Indiana and Michigan and the History of the Great Lakes* in 1915, co-authored with Frank Leverett with whom he worked in the US Geological Survey (USGS). His views on glaciation would be adapted to continental drift.

[20] Celal Sengor, "Eduard Suess and the Origin of Modern Geology", pp. 16–17.

[21] Frank Bursley Taylor, "Bearing of the Tertiary Mountain Belt on the Origin of the Earth's Plan" in *Bulletin of the Geological Society of America*, Vol 21, pp. 179–226, Pl. 4, June 3, 1910.

[22] Alex L. Du Toit, *A Geological Comparison of South America with South Africa* (Washington, Carnegie Institution of Washington, 1927), p. 2.

This work demonstrated that his practical experience with the USGS had compensated for his lack of formal education in geology.[23]

Taylor also drew upon his earlier studies in astronomy in order to formulate an early theory of continental movement. While the paper (self-published in 1898) was rather ambitious in challenging the work of Isaac Newton, it is perhaps best noted for his use of the role of the gravitational pull of the moon in shifting land masses away from the poles towards the equator. He would refine this theory, drawing upon the work of Eduard Suess in his paper on continental drift for which he is most remembered.[24]

In this seminal paper, "Bearing of the Tertiary Mountain Belt on the Origin of the Earth's Plan", Taylor begins by describing a global mountain chain that appears to straddle the earth, including the oceans, where it appears above sea level in the form of island groups. This mountain chain apparently originated in the Tertiary period (a vast period of time covering 66 million to approximately 2.5 million years ago). For Taylor, this epoch of mountain-building required a suitable explanation which is the main thread running through his paper.[25] Prior to plate tectonics, the theory of mountain-building or orogeny had been largely based on the work of James Hall in the mid-nineteenth century, derived from his understanding of the Appalachian Mountains and following the prevailing view of uniformitarianism that emphasised sedimentation and deposition. According to Hall, the Appalachian mountain range was a product of the progressive accumulation of sediment. The weight of these sedimentary deposits caused the underlying crust to sink. The sinking crust then created the conditions by which the overlying sedimentary rock was deformed and, in some areas, experienced the intrusion of magma from the mantle which introduced igneous rock into the formations. The area subsequently experienced uplift which was then subject to erosion. For Hall, then, orogeny could be explained through the observable processes of deposition and subsidence. James Dwight Dana argued instead that, as the earth cooled, it contracted and this had a similar effect on the crust

[23] Michele L. Aldrich, "Taylor, Frank Bursley" in *Complete Dictionary of Scientific Biography* (Charles Scribner's Sons: 2008) taken from http://www.encyclopedia.com/people/science-and-technology/geology-and-oceanography-biographies/frank-bursley-taylor Accessed 7 October 2016. For further information, a more detailed biography of Frank Bursley Taylor is Frank Leverett "Memorial to Frank Bursley Taylor", *Proceedings of the Geological Society of America for 1938* (1939), pp. 191–200.

[24] Aldrich, "Taylor, Frank Bursley".

[25] Taylor, "Bearing of the Tertiary Mountain Belt", pp. 179–180.

and most likely occurred on the borders between continental and oceanic crust. The crust was then deformed, forming geosynclines which were large concave areas that accumulated sediment and geanticlines which were raised areas subject to erosion.[26] Both explanations, however, appear rather limited when applied to the vast Tertiary mountain range under discussion and Taylor contended that this mountain formation could have possibly arisen as a result of processes occurring on the continental land masses on which they were found rather than as a result of oceanic processes.[27]

At the same time, this mountain range could highlight the processes that underlay orogeny. This was due both to its extent and its relative youth which meant that it retained its height and had not as yet been subject to the ravages of erosion. At this point, Taylor began heavily to rely on the work of Eduard Suess, adapting, refining and criticising Suess's hypotheses in order to develop his own formulation. Suess, too, had concluded that the Earth had cooled and subsequently shrunk. This had led to the crust being drawn inwards—the parts that subsided most formed the ocean depths and those that had experienced the least subsidence formed the land masses. This differential height allowed for the accumulation of sediment which was then contorted, forming new land mass and folded into mountain ranges. Suess' interpretation had come from his study of the mountain ranges of Asia where this simple movement from north to south could most clearly be seen. Taylor believed that this view could be adapted to the planet as a whole.[28]

Whereas Suess had confined his focus to Siberia as the centre from which crustal expansion occurred, Taylor was more ambitious suggesting movement of the earth's crust on a much larger scale. Suess had also considered crustal movement as moving southwards from northern latitudes. Taylor found support for this in his own discussion of the Indian subcontinent as well as the separation of Greenland from North America and Eurasia but further refined this by considering whether crustal movement in general occurred from "high latitudes toward low latitudes". To support his hypothesis, he would need to turn his focus on the southern

[26] Harold Williams, "Introduction" in *Geology of the Appalachian-Caledonian Orogen in Canada and Greenland*, Harold Williams, (ed) (Geological Society of America and Geological Survey of Canada, 1994), pp. 12–13.

[27] Taylor, "Bearing of the Tertiary Mountain Belt", p. 180.

[28] Taylor, "Bearing of the Tertiary Mountain Belt", pp. 181–182, 183, 186, 188–189.

hemisphere. By hypothesising the movements of the southern hemisphere landmasses, that is, Africa, South America, Australia and the Malay Archipelago, he speculated that the movement of the crust in the southern hemisphere occurred in the opposite direction to that in the north viz. from south to northwards. This appeared to bear out his initial view of crustal movement from "high latitude" or the pole towards the equator.[29]

The southern hemisphere arguably also provides the most visible evidence for continental drift and Taylor was highly cognisant of this. In addition to the ubiquitous image of the fit between the west coast of Africa and the east coast of South America, Taylor highlighted the importance of the Mid-Atlantic Ridge and its location midway between the two continents. His visionary explanation would only be substantiated half a century later:

> It is probably much nearer the truth to suppose the mid-Atlantic ridge has remained unmoved, while the two continents on opposite sides of it have crept away in nearly parallel and opposite directions.[30]

For Taylor, as with Wegener who followed him, the evidence for continental drift was compelling—it was an explanation that best fit observations made on a global scale. But, if the crust did move or creep or drift, what then caused it to do so? Taylor envisaged nothing less than a change in the shape of the earth. Once again, Suess came into play with his notion of "polar flattening". The earth was once considered to have been a perfect sphere but the effect of flattening at the poles transformed it into the "oblate spheroid" evident today. The change in the shape of the earth combined with the preponderance of land mass in the northern hemisphere and ocean south of the equator suggested to Taylor that continental movement was due to a shifting centre of gravity. Movement would begin in the north, shifting the centre of gravity to the south. This would then lead to a corresponding movement in the south and the pattern would continue *ad infinitum*, all leading to crustal movement towards the equator. This also explained the concentration of land in the north, the site of the first initial crustal movement and growth.[31]

[29]Taylor, "Bearing of the Tertiary Mountain Belt", pp. 191, 204, 218.
[30]Taylor, "Bearing of the Tertiary Mountain Belt", p. 218.
[31]Taylor, "Bearing of the Tertiary Mountain Belt", pp. 220–221.

Taylor's explanation for continental drift would share an important feature with that of Alfred Wegener. While the evidence for continental drift was exhaustingly spelt out, the mechanism of the drift itself was relegated to the realm of supposition—which would be the point of weakness which opponents of drift theory attacked, reserving a great deal of their vehemence for the man most associated with continental drift.

Alfred Wegener

There are two images of science that exist in the popular consciousness—one is the evolutionary development of humankind. It is a simplified linear progression, ignoring the web-like complexity and myriad evolutionary dead ends. The other is the obvious jigsaw-puzzle-like fit between the west coast of Africa and the east coast of South America. Both images suggest the centrality of Africa—something that Smuts was eager to capitalise on. In 1925, Jan Smuts addressed the South African Association for the Advancement of Science—of which he was president—where he emphasised Africa's (and South Africa's) significance as a site of human origins as well as its geological significance as part of Gondwana, embracing Wegener and Du Toit's work. At the time of his speaking, debates raged over humankind origins and drift theory was hardly mainstream geological theory, yet they fit into Smuts's ideological framework.[32] It was in the eastern and southern regions of the African continent that the unearthing of fossils of early hominins provided convincing evidence that humankind's origins lay in Africa. The other image suggested the fit between landmasses now separated by thousands of miles of ocean. There is a significant difference between the two, however. The evolutionary chain of progression is a pictorial depiction illustrating Darwin's theory whereas the perceived fit between Africa and South America required explanation—which Alfred Wegener would attempt to provide.

Wegener enrolled at the Friedrich-Wilhelms University in Berlin to study astronomy. He would also spend a semester at the Ruprecht-Karls University in Heidelberg studying meteorology which was, at the time, considered part of the same discipline as astronomy. Wegener and his brother, Kurt, subsequently enrolled for a semester at the University of Innsbruck in Austria. The university was dominated by the Eastern Alps

[32] Saul Dubow, "Global Science, National Horizons: South Africa in Deep Time and Space" in *The Historical Journal*, Cambridge University Press, 2020, pp. 11–12.

and the Wegeners studied botany and field geology here. Of particular interest in their introduction to alpine botany was the relationship between plant types and their location in terms of latitude. This was also a particularly significant time for the study of geology as well. Since the late nineteenth century the focus had been on alpine geology, the clearly intense and large-scale processes that had created the dramatic alpine landscape of folds and deformation.[33]

While he was awarded his PhD in astronomy from the University of Berlin, Wegener found a career in meteorology to be more to his taste and joined a Danish expedition to Greenland where he was tasked with collecting atmospheric data for two years using weather kites and balloons.[34] The Danmark Expedition was one of several to the northern latitudes from the late nineteenth century but was touted as "the largest polar-scientific endeavour ever mounted" with a contingent of 28 scientists, including Wegener who was tasked with physics and meteorology. It was on this first trip that Wegener also carried out geological work—learning how to collect rock and fossil samples that would ultimately be used to create a geological map of Greenland. He became particularly adept at this as well as photography where he was able to capitalise on his "strongly visual imagination and intuition" which would stand him in good stead later on.[35] Upon his return to Germany two years later, Wegener took up a position at the University of Marburg. While his area of expertise was astronomy and meteorology, he soon turned his attention to geology.[36]

In 1910, Wegener was struck by the almost perfect fit between the eastern coastline of South America with the western coastline of Africa. Although Wegener noticed the fit between the coastlines of South America and Africa and its implications, he was certainly not the first to do so. As early as the late sixteenth century, Abraham Ortelius, a Dutch geographer, implied that the complementary nature of the coastlines suggested that

[33] Mott T. Greene, *Alfred Wegener: Science, Exploration, and the Theory of Continental Drift* (Baltimore, Johns Hopkins University Press, 2015), pp. 12–15, 17–19, 21, 27–28, 34–36.

[34] Lisa Yount, *Alfred Wegener: Creator of the Continental Drift Theory* (New York, Chelsea House, 2009), pp. 20–26.

[35] Greene, *Alfred Wegener*, pp. 93, 135–136.

[36] Yount, *Alfred Wegener: Creator of the Continental Drift Theory*, pp. 27–29.

the two continents had once been united.[37] Ortelius is an illustration of the difficulty in determining a point of origin for scientific theory—notions of continental drift could feasibly be traced far earlier than the nineteenth century. The European voyages of discovery, beginning in the late fifteenth century, promoted folklore and tall tales of "lost continents". Ortelius incorporated these into his map of the world *Typus Orbis Terrarium* in 1570. In this map, Ortelius had drawn large hitherto unknown continents lying on both the north and south poles, the latter pre-empting Antarctica. He also depicted this southern continent as being linked to South America. His cartographic depiction had its origins almost 1500 years earlier when the Greek, Hipparchos of Rhodes theorised a large southern landmass that would provide a "balance" to the continents of the northern hemisphere. Ortelius, however, was the first to suggest that the fit between the coastlines of Africa and South America was due to the separation between the two, with the Atlantic Ocean taking its place between them. This separation had been caused by some calamitous event that had rent this large continent asunder. In his supposition, Ortelius was supported by the eminent Sir Francis Bacon. Alexander von Humboldt in the early nineteenth century also proposed the possible way in which South America and Africa were joined however the significance of Ortelius' work was lost to history until as late as 1994 when his far-sighted hypothesis was resurrected as part of the history of continental drift.[38] It was, however, Alfred Wegener that was able to lend scientific weight to this conjecture.[39]

In the introduction to his work, however, it appeared as if Wegener had rather haphazardly stumbled onto continental drift—a theory that appeared to be based on a surfeit of speculation. According to Wegener, the seeds of continental drift took hold as early as 1910 when he happened to notice the fit between the continents separated by the Atlantic Ocean as he was "considering the map of the world". A year later, he "came quite accidentally upon" a publication that described the fossil similarities between Africa and Brazil underpinning the notion of "a former land bridge" between the two. This inspired him to "[undertake] a cursory

[37] M.J. De Wit, B.B. De Brito Neves, R.A.J. Trouw and R.J. Pankhurst, "Pre-Cenozoic Correlations across the South Atlantic Region: 'the ties that bind'" in *West Gondwana: Pre-Cenozoic Correlations Across the South Atlantic Region* (London, The Geological Society of London, 2008), p. 1. Downloaded from http://sp.lyellcollection.org. Accessed 2 October, 2016.

[38] Nield, *Supercontinent*, pp. 27–29.

[39] De Wit et al. "Pre-Cenozoic Correlations across the South Atlantic Region"

examination of relevant research in the fields of geology and palaeontology" which lent "weighty corroboration" and culminated in his theory of continental movement.[40] Perhaps written in a modest tone, what appeared to be a lack of scientific rigour at the very outset of his proposition seemed almost designed to arouse the ire of his detractors.

However, Wegener's observation of the complementary nature of the coastlines of South America and Africa was not as superficial as his detractors supposed. It was less about the fit between coastlines—which varied tremendously over time—than it was of the continental shelves. The Challenger Expedition was a British research expedition, the first to engage solely in oceanographic scientific research with a focus on the chemical composition of sea water, currents, marine biology and geology. A Navy warship was transformed into a floating laboratory and the expedition was led by scientists John Murray and Charles Wyville Thompson. Over its four-year voyage the ship traversed the Indian, Atlantic and Pacific Oceans, discovering the Marianas Trench and what would become known as the Mid-Atlantic Ridge—which would play a central role in plate tectonics theory.[41] The Challenger Expedition had, between 1873 and 1876, taken depth readings of the Atlantic Ocean which had helped reconstruct the continental shelves of South America and Africa and these were published in the *Allgemeine Handatlas* by Karl Andree—which was where Wegener had made his observation.[42]

Based on the research questions posed by an earlier Norwegian explorer, Fridtjof Nansen, Wegener and his friend, Johan Peter Koch who had accompanied him on the Danmark Expedition, proposed a trip to Eastern Greenland that would investigate glaciation and the geography of the terrain. Of particular interest was Nansen's discussion of the evidence of tropical flora in Greenland which suggested that the Earth's rotation may have been very different in the past. Nansen, himself, had suggested "the migration of the Earth's pole of rotation" as an explanation for what were clearly very different palaeoclimates. While undertaking research into the proposal for the Greenland expedition, a serendipitous moment occurred when Wegner came upon an article published in the new German

[40] Alfred Wegener, *The Origins of Continents and Oceans*, John Biram, (translator) (New York, Dover Publications Inc, 1966), p. 1.

[41] Woods Hole Oceanographic Institution, "Dive and Discover: Expeditions to the Seafloor", https://divediscover.whoi.edu/history-of-oceanography/the-challenger-expedition/ Accessed 13 June 2020.

[42] Greene, *Alfred Wegener*, pp. 214–216.

geological journal *Geologische Rundschau*. Written by Erich Krenkel, the article suggested the prior union of Brazil with Africa as evident by the discovery of fossilised clams. These landmasses were subsequently separated by an ocean, leading to later biological diversification between the two continents. This substantiated Wegener's own visual instinct when observing the continental shelves of South America and Africa. A further article by Konrad Keilhack drew Wegener's attention to the prevalence of *Glossopteris* across India, Africa and Australia. The article also fired Wegener's enthusiasm by positing similar glacial activity across these landmasses. This had the potential to explain the anomalous botanical fossil finds of Greenland.[43]

For Wegener, the explanation was that the continents had moved or been "displaced". He embarked upon feverish activity, writing and revising his hypothesis in 1911. While criticised for his lack of fieldwork, Wegener's work followed in the German tradition of "syntheses". He was able to draw upon an impressive array of publications—much of it fairly recent—in the fields of geology, palaeontology and physics in order to substantiate his hypothesis, including the earlier work of Eduard Suess. Of particular use was *Evolutionary Development of the Continents and their Life Forms* by Thedor Arldt which summarised the distribution of fossils across continents, showing their similarities across landmasses now far apart.[44]

Just a few months later, at the beginning of January 1912, Wegener presented his ambitious hypothesis to the Geological Association in Frankfurt am Main. The title of his paper was "The Geophysical Basis of the Evolution of the Large-Scale Features of the Earth's Crust (Continents and Oceans)". A subsequent paper, "Horizontal Displacements of the Continents", was presented four days later to the Society for the Advancement of Natural Sciences.[45] In these papers, Wegener proposed nothing less than a complete change in the way in which the geological processes of the Earth were understood.

The prevailing geological view explaining the fit between continents—as well as the similar fossils found in West Africa and Brazil and other resemblances across distinct landmasses—was that "land bridges" had existed joining land masses together and these had subsequently sunk into the ocean, leaving little trace. Wegener's own understanding of geophysics

[43] Greene, *Alfred Wegener*, pp. 220–223, 233–235.
[44] Greene, *Alfred Wegener*, pp. 241–245.
[45] Wegener, *The Origins of Continents and Oceans*, p. 1.

suggested that the "land bridge" hypothesis was implausible and, in 1915, he published *Die Enstchung der Kontinent un Ozeane* (*The Origin of Continents and Oceans*) where he argued that continental landmasses had been in a state of motion throughout the earth's history—their locations, forms and interaction with each other had been in a state of flux. Thus, at some prehistoric point in the history of the earth, the continents of South America and Africa had been joined and had subsequently broken apart.[46]

This movement, translated into English as "continental drift" was, from the outset, mired in controversy. A geological outsider had had the temerity to challenge two existing hypotheses—that of the "land bridges" and another view that the earth's geology was a result of the earth's cooling since its creation and subsequent contraction, making geological features such as mountains "wrinkles" on the earth's surface.[47]

In contrast, Wegener proposed that there had been initially a single land mass that he termed Pangaea ("all-Earth") which had been surrounded by an ocean Panthalassa ("all-sea"). This giant continent had then broken apart and its constituent pieces had begun drifting. Wegener drew upon a formidable range of evidence to support this hypothesis including the similarities between rock formations and fossilised flora and fauna across continents. He further made use of geophysics to indicate that there were two forms of crust—less dense granitic crust characteristic of land masses and denser basaltic oceanic crust. He also brought to bear his training in meteorology—the evidence of tillite in temperate climates indicated previous glaciation and the discovery of coal seams in colder latitudes suggested that they had once harboured lush tropical vegetation. This was further evidence of continental movement.[48]

Wegener's work was extensively rewritten and revised in a number of editions, indicative of his response to the controversy it aroused, what Mott Greene in his extensive biography of Wegener refers to as "who had the right to speak".[49] Wegener's hypothesis was not based on the conventional methods used by contemporary geologists and, as a meteorologist, he was an outsider. Further, his theory had a strong basis in physics, the branch of science in which he had been extensively trained. While both physics and chemistry were minor components of geology in

[46] Yount, *Alfred Wegener: Creator of the Continental Drift Theory*, pp. 33, 38–39.
[47] Yount, *Alfred Wegener: Creator of the Continental Drift Theory*, p. 39.
[48] Yount, *Alfred Wegener: Creator of the Continental Drift Theory*, pp. 41, 46–51, 54–55.
[49] Greene, *Alfred Wegener*, pp. 248, 254.

the early twentieth century, it was a science that was based on practical experience and fieldwork. Wegener's work referred to processes that had to be inferred.

Wegener's initial hypothesis of displacement appeared to be a middle ground between two very different views of geology. The Europeans emphasised the notion of a cooling, contracting Earth, allowing for the creation of continents and oceans which were a product of "wrinkles" on the Earth's surface. American geologists, on the other hand, believed that both oceans and continents were permanent features, neither created nor destroyed. Wegener also used the principle of isostasy to refine his notion of displacement.[50] Isostasy relates to the equilibrium achieved by the earth's crust or lithosphere as it lies atop the mantle, for instance, continental crust is less dense than oceanic crust so "floats", lying at a higher elevation above the mantle. This level may change depending on the changing density of the material—new mountain chains lie lower due to their greater density but subsequently rise, as they are subject to the denuding actions of erosion.[51]

Wegener underestimated the extent of criticism his work would cause. He believed that his explanations, based on understood principles of physics, would be compelling for geologists largely unfamiliar with physics. In this he was mistaken. Wegener's understanding of the physics of motion was further enhanced in a subsequent expedition to Greenland in 1912. Although harrowing, filled with incident, and largely unsuccessful, the trip presented Wegener with the opportunity to study the Jacobshavn Isbrae, a fast-moving glacier that travelled between 15 and 30 metres daily. It was a small leap from watching the flow of solid ice to hypothesising a similar flow of solid rock.[52]

While recuperating from injury during the First World War and diagnosed with "a chronic heart ailment", Wegener worked on yet another revision. In this he was assisted by another geologist, Hans Cloos, who helped Wegener navigate the tricky terrain of geology. This edition, published in 1915, situated Wegener's work within an existing narrative of the development of geology. He demonstrated the early origins about debates regarding floral and faunal similarities across distinct continents

[50] Greene, *Alfred Wegener*, pp. 248, 254.

[51] Brian J. Skinner, Stephen C. Porter and Jeffrey Park, *Dynamic Earth* (Hoboken, John Wiley and Sons, 2013), pp. 34–35.

[52] Greene, *Alfred Wegener*, pp. 256, 312.

which could be traced to the beginnings of modern geology. He also more assertively proposed that the weaknesses of the American and European views of the Earth could be addressed if "lateral displacement" of continents were accepted. The following edition of his work, published in 1920, included a visual depiction of what the continents had looked like 250 million years ago. Here, he included a feature called the "Lemurian Compression" which would be a result of India's collision with Europe.[53] His choice of name, a reference to the lost mythological continent of Lemuria, demonstrated the hold that myth retained in science.

Despite his many revisions, Wegener could not adequately account for the mechanisms that drove continental drift—the term "drift" itself was vague—and this was seized upon by his detractors, of whom there were many.[54] By the 1920 edition, Wegener had incorporated Frank Bursley Taylor's explanation for movement.[55] Wegener's vague explanation for continental movement as possibly being a result of the centrifugal force produced by the earth's rotation provided fuel for the fire yet Wegener had tantalisingly considered the importance of the Mid-Atlantic Ridge as the site where new crust was continuously created.[56] This was an approach that Du Toit would develop further in his own elucidation for continental movement. Wegener was later subjected to a fresh round of criticism when his work was finally published in English in 1924. There were, however, varying levels of criticism—American geologists were particularly hostile as their scientific methodology relied on inductive reasoning: the marshalling of evidence which would ultimately lead to a theory to account for it. European geologists, on the other hand, provided for its opposite—deductive reasoning—where evidence could be collected in support of an existing theory.[57] Wegener had little patience for the dismissal of his work as largely theoretical with insufficient consideration of facts as evident in his response to criticism from zoologist, Georg Pfeffer: "He belongs to that crowd of people who boast about standing on a solid ground of facts and having nothing to do with hypotheses, never realizing that their solid ground of facts has embedded in it a completely false hypothesis!"[58]

[53] Greene, *Alfred Wegener*, pp. 326, 327, 329–331, 414.

[54] Yount, *Alfred Wegener: Creator of the Continental Drift Theory*, p. 56.

[55] Greene, *Alfred Wegener*, p. 421.

[56] David Whitehouse, *Journey to the Centre of the Earth: A Scientific Exploration into the Heart of Our Planet* (London, Weidenfeld and Nicholson, 2016), p. 88.

[57] Yount, *Alfred Wegener: Creator of the Continental Drift Theory*, pp. 62–63, 66–67.

[58] Greene, *Alfred Wegener*, p. 374.

Further criticism was because Wegener's "continental drift" smacked of biblical catastrophism which had been discarded in favour of uniformitarianism and the acceptance of evolutionary theory.[59] The acrimony created by Wegener's theory is, therefore, an illustration of the context in which scientific knowledge is produced; existing assumptions create a complexity that belies the scientific method and the use of reason.

The final and fourth edition of Wegener's work was published in 1928 in the wake of the research undertaken by Alex du Toit in South America. In the previous edition—the first to be translated into English and arguably the most popular, Wegener had already addressed the glaciation of South Africa, suggesting that, based on the glaciation pattern, South Africa was centrally placed with the other continents situated around it and was where glaciation had its origins. By 1928, he was able to draw upon Du Toit's geological evidence in response to his critics and, according to Greene, "Wegener recognized Du Toit's book immediately for what it was: the most influential geological support he had ever received".[60]

At the same time, Taylor believed that Wegener had also been influenced by his research. In a letter to Du Toit, just prior to the publication of *Our Wandering Continents* in 1937, Taylor believed that it was his paper "Bearing of the Tertiary Mountain Belt on the Origin of the Earth's Plan" presented to the Geological Society of America (GSA) at the end of 1908 that marked the first attempt to systematically provide evidence for continental drift. A delay in the printing of the paper in the Bulletin published by the GSA meant that it only appeared in print in 1910. It was, however, noticed by Wegener who made brief reference to it in a German publication in 1911. Wegener's subsequent work on drift was therefore believed by Taylor due to the inspiration provided by his own paper. Yet Taylor also disagreed with some of Wegener's suppositions such as the "wandering of the poles" which he considered to be too ambitious.[61]

Du Toit's response to Taylor's letter was a model of diplomacy as he thanked Taylor for his input to the history of the theory. Du Toit also claimed to have been similarly inspired by a reference to Wegener's work before having had the opportunity to actually read Wegener's work. Yet

[59] Yount, *Alfred Wegener: Creator of the Continental Drift Theory*, p. 68.

[60] Greene, *Alfred Wegener*, pp. 455, 462, 546, 548.

[61] UCT-JL: Alex L. Du Toit Papers—BC 722: C1—Research Papers Relating to his Book "Our Wandering Continents": C1.1 Correspondence (1935–1938) Letter by FB Taylor to Du Toit, 5 February 1937.

Du Toit was less enamoured with the historical record as Taylor was "Anyhow nothing much would seem to be gained by probing into the matter now".[62] However, in the history of continental drift addressed in *Our Wandering Continents* he was careful to give both Taylor and Wegener their due acknowledgement. The incident symbolises the convergence of thinking on the subject as well as the vagaries of historical memory—it would be Wegener not Taylor who would be most associated with continental drift.

Alfred Wegener would never live to see his theory vindicated—he died on the Greenland ice of a heart attack in 1930 on yet another expedition, decades before "continental drift" or plate tectonics entered mainstream geology.[63] Wegener believed that drift between Europe and Greenland could be most easily observed. He would use the latitude measurements taken on previous explorations in the nineteenth century and triangulate them with those taken by J.P. Koch, a cartographer on his expedition. The calculated results suggested a rather head-whipping speed of between 9 and 32 metres per annum when the common image used to illustrate the rate of continental movement is that of fingernail growth—for geologists, still a fair rate. Wegener's measurements were hampered by contemporary technology and it is only over the past few decades and the use of Global Positioning Satellites that accurate measurements are possible.[64] As a postscript, the remainder of the research undertaken in Greenland—under the leadership of Kurt Wegener—did provide support for isostasy.

With the unrelenting criticism that greeted the work of Wegener and Taylor, it is therefore all the more remarkable that Wegener's work found a receptive audience in South Africa with Alexander Du Toit its most outspoken proponent and Jan Smuts's advocacy of the hypothesis.

[62] UCT-JL: Alex L. Du Toit Papers—BC 722. C1—Research Papers Relating to his Book "Our Wandering Continents": C1.1 Correspondence (1935–1938) Letter by Du Toit to FB Taylor, 20 March 1937.

[63] Yount, *Alfred Wegener: Creator of the Continental Drift Theory*, p. 85.

[64] Nield, *Supercontinent*, pp. 151–152.

CHAPTER 6

Looking Through … the Keyhole of Nature: Du Toit and Early Continental Drift

The South African Association for the Advancement of Science (SAAAS) held its nineteenth annual meeting in Durban, Natal, in July 1921. Flanked by a description of tours of the "Indian market, Municipal Native eating house and Native brewery" and "a trip to the Bay" and "zoological excursion in the afternoon", the Report of the Nineteenth Annual Meeting of the SAAAS contained the following:

> At 8.15 p.m., Dr A L Du Toit, BA, FGS, gave a popular illustrated lecture on "Land Connections between the other Continents and South Africa in the Past," in the Arthur Smith Hall, the President of the Association presiding.[1]

This public lecture, delivered on 15 July 1921, contained the key points that Du Toit would use to make his case for continental drift and laid the foundation for the seminal paper that was published three months later. Written with a minimal of geological jargon, the lecture begins by

---

The quote appearing in the title is accredited to Jacques Cousteau in *Christian Science Monitor*, 21 July 1971.

[1] "Report of the Nineteenth Annual Meeting of the South African Association for the Advancement of Science," in *The South African Journal of Science*, Vol. XVIII (Johannesburg, South African Association for the Advancement of Science, 1922).

S. Chetty, *'Africa Forms the Key'*, Cambridge Imperial and Post-Colonial Studies, https://doi.org/10.1007/978-3-030-52711-2_6

introducing the basic elements upon which his proposition was based, the first being his use of biostratigraphy where the unearthing of fossil types could be correlated with similar fossil finds all over the world, suggesting that they had been deposited in the same period. Another key principle was his use of evolution and natural selection. Du Toit explained simply that evolutionary divergence was more apparent on land than it was in the oceans and took place relatively quickly due to geographical and other constraints that favoured natural selection. The corollary of this was that, where species were very similar, the argument could be made that they shared a similar geographic and climatic environment. Simultaneously—and here he took particular issue with the view of "land bridges" that dominated—the possible existence of bridges linking landmasses did not presuppose the migration and similarity of fauna and flora across these disparate regions. On the contrary, the overwhelming weight of biological evidence suggested instead that these landmasses were more closely connected in the past. Du Toit situated his work within a scientific pantheon of the late nineteenth century that included Thomas Huxley and Eduard Suess. While no mention was made of Frank Bursley Taylor, he also briefly alluded to the contemporary work of Alfred Wegener.[2] The evidence for Du Toit's argument, however, was drawn from the area with which he was most familiar—glaciation.

In addition to the dramatic geological events such as earthquakes and volcanoes, every geologist should see a glacier. While these vast, moving accumulations of ice lack the pyrotechnic splendour of volcanoes or the incredibly fast destructive potential of earthquakes and their very name is an adverb for sluggishness, glaciers are one of the most important means by which landscapes are shaped. They are formed at high latitudes or on mountain tops where there is insufficient heat in summer to fully melt winter's snowfall. In this way, the snow begins to accumulate and the weight of it leads to the creation of ice. As the ice grows steadily with every successive winter, there is eventually a sufficient amount of it to be subject to gravity and it begins to move. There are two forms of glaciers—valley and continental glaciers. The latter is also described as ice sheets and it is to this that Du Toit refers in his discussion of Gondwana and, at present, Antarctica remains covered by a large sheet. While glaciers usually enter

[2] Alex L. Du Toit, "Land Connections between the Other Continents and South Africa in the Past," Public Lecture given on 15 July 1921, pp. 120–122, 136. Published in *South African Journal of Science*, 18, 120–140.

popular consciousness through their increasing diminution—a visible signifier of global warming—they are the most effective means of erosion, tearing off surrounding rock and transporting them great distances. The rocks that lie close to the bottom of glaciers are scraped along the underlying bedrock. They leave behind deep scars termed striations. These striations thus allow geologists to trace the movement of the glacier. The debris carried by glaciers are referred to as moraines and are left behind as the glacier retreats. This sediment consisting of poorly sorted sand and rock is till. Over time, this sediment accumulates becoming rock known as tillite (Fig. 6.1).[3]

Du Toit drew attention to the very distinct climates of the northern and southern hemispheres. The Carboniferous Period was named for the large forested expanses of the northern hemisphere, the carbonised remains of which were responsible for extensive coal deposits that are still exploited today. The southern hemisphere, on the other hand, was rather chillier but gave way to the milder Permian which fostered the growth of the hardier *Glossopteris*, the remains of which were also partly responsible for the coal deposits found south of the equator.[4]

In this lecture Du Toit's argument rests on three points—glaciation, palaeobotany and volcanism. After acquainting his audience with glaciation as a key geological process, Du Toit discussed the evidence for it in South Africa, highlighting Natal where lithified moraine could be found in the macadam of Durban's roads and at the Dwyka River in Prince Albert in the Western Cape, from which it received the name, the Dwyka Conglomerate. Based on this glacial evidence Du Toit showed that the glacier began in the Transvaal, spreading into present-day Namibia, the Cape and further east. By comparing the Dwyka Conglomerate with South America, the Falklands Islands and parts of India and Australia, Du Toit highlighted that the glaciation occurred across these diverse and distant landmasses during the same time period. Moreover, only parts of these areas had experienced glaciation. Du Toit's original—and what he believed to be controversial—perspective here is that, taking into account the pattern of the glaciation which suggested the orientation of the landmasses involved, Gondwana was smaller than had previously been supposed, was in closer proximity to the South Pole (to account for the

[3] John Grotzinger and Thomas H. Jordan, *Understanding Earth (Sixth Edition)* (New York: W.H. Freeman and Company, 2010), pp. 573–574, 572, 580–582, 585, 592.

[4] Du Toit, "Land Connections," p. 123.

Fig. 6.1 Glaciated rock, indicating striations (University of Cape Town Libraries, mss_bc722_c2_4_001)

glaciation) and had been covered by a single ice sheet during the Carboniferous Period. Contemporary land masses were therefore the remnants of this continent, their coastlines altered to an extent in the intervening time period by geological processes. For Du Toit, this accounted for the geological and fossil evidence drawn from these regions.[5]

With regards to fossil evidence, in addition to the ubiquitous *Glossopertis*, Du Toit also spoke of the prehistoric reptile *Mesosaurus*, the fossilised remains of which were found embedded in the rock strata of continents separated by vast ocean. *Glossopteris* is, of course, highly significant and Du Toit traced its migration with its very success suggesting that there was little obstacle (such as ocean) to its movement. Only later, with the

[5] Du Toit, "Land Connections," pp. 124–126.

migration of flora from the northern hemisphere, was *Glossopteris* usurped. The hardy *Glossopteris*, which required little water, was not perturbed by the cold climate of the southern hemisphere and Du Toit proposed that, as it spread, grazing animals followed in its wake. In the Late Triassic period, however, the fossil evidence began to change suggesting that Gondwana had connected to Europe. As a result, there was an increasing dominance of northern flora and fauna. This migration was, however, not unidirectional and there was some evidence of particularly South Africa faunal fossil evidence found in Scottish sandstone.[6]

Fossil evidence may be considered biostratigraphic and Du Toit then turned his attention to lithostratigraphy or the rock itself. The sandstone and lack of plant fossils evident in the Bushveld Sandstone and Cave Sandstone in South Africa indicate that, more than two hundred million years ago, the climate during the Triassic period had grown increasingly dry; similar records of climatic change was preserved in the rock of Zambia, the Congo and India. It was at this point that the landmasses that comprise South Africa, South America and the subcontinent of India experienced tumultuous volcanic upheaval as a prelude to the fragmentation of Gondwana. This volcanism was indicated by the basalts formed by solidified lava erupting at the surface and by dolerite, formed by magma cooling and solidifying before it reaches the surface. These eruptions had a detrimental effect on the fossil record, disrupting the deposition of sediment. This meant that fossil evidence was largely only obtainable from coastal regions. The lithographic evidence for this period was thus further illustration for the case for Gondwana and continental drift.[7]

However, while it was important to introduce the concept of continental movement, that was only part of the story. Just as important was the need for an explanation as to why it occurred. In this lecture, Du Toit attempted an explanation for both orogeny (mountain-building) and continental drift. To the modern reader, the account of orogeny appears somewhat simplistic—or that is perhaps a result of how Du Toit explained it to his audience. The subsidence of parts of the earth was due to the weight of sediment deposited. As these areas sank, this caused the adjacent crust to rise. Sediment is usually deposited from land into the ocean which leads to sinking oceans whereas land is elevated. At the same time, the Earth's crust is also subject to what Du Toit described as "lateral pressure"

[6] Du Toit, "Land Connections," pp. 127–128.
[7] Du Toit, "Land Connections," pp. 130–131.

as a result of "creeping substratum" or the movement of the mantle. The combination of these events, where uplifted crust is squeezed laterally, results in folds, that is, mountain chains as well as volcanism. This could be seen in South Africa where the massive weight of extensive sedimentation led to "collapse" of the crust with the resulting formation of the Southern Cape Ranges—and its effects could also be seen in South America as well as the Falklands Islands. This process of orogeny reached its zenith during the Jurassic period by which point Gondwana had likely split into two—the landmass comprising Australia and Antarctica and another comprising South America, Africa and India. Madagascar then parted ways with Africa and, following that, India. This was marked by further volcanism as evident in the famous Deccan Traps of central India and, as Du Toit hypothesised, the diamond-bearing kimberlite pipes that would have such a powerful impact on the history of modern South Africa.[8]

For Du Toit, the catastrophic break-up of Gondwana (and his explanation for continental movement) was due to the "tear-lines" or areas of weakness in the earth's crust that were exploited by "centrifugal forces set up through spinning around the polar axis". He suggested that the "spinning" had the effect of widening these weak zones or cracks, eventually leading to complete separation. The gaps would then be filled by ocean and the land masses would drift further apart. As they moved, they exerted pressure ahead of them and the subsequent formation pushed up mountains. According to Du Toit this movement was not "hypothetical" but could be seen by the increasing distance between the United States and Britain which he believed to be "several yards per annum"—which we now know to be entirely too fast.[9] However, while Du Toit's evidence made a compelling case for Gondwana and its subsequent fragmentation, his explanation still remained firmly in the realm of the "hypothetical".

From the hypothetical he concluded with the mythical, speculating that Atlantis may have been a remnant of the connection between South America and North Africa, which had subsequently subsided into the ocean which bears its name.[10] This was a rather romantic end to a lecture that had begun by using undisputed principles in evolutionary biology and stratigraphy to make a persuasive argument. It was also indicative of the way in which mythological images of lost lands still retained a hold on the

[8] Du Toit, "Land Connections," pp. 132–134.
[9] Du Toit, "Land Connections," pp. 135–136.
[10] Du Toit, "Land Connections," p. 140.

scientific mind. This lecture though, while aimed at a popular audience, would merely serve as an introduction to the scientific paper that followed in October—which established Du Toit as a prescient figure in continental drift theory.

"The Carboniferous Glaciation of South Africa" was published in October 1921 and was quite unlike the public lecture that Du Toit had delivered in July. Aimed at the professional geologist rather than a general, albeit scientifically minded audience, Du Toit adopted the empirical and descriptive style of writing that tended to characterise his academic work. And, unlike the lecture which drew upon the accumulation of a variety of evidence to make the case for continental drift, here Du Toit restricted himself to glaciation in southern Africa, the field of study in which he arguably had the most experience. While Du Toit made the modest claim that there was still much that remains incomplete in the geological record, by emphasising the substantial fieldwork that he had carried out in the region for almost two decades, he was able to lend his claims an air of authority.[11]

Simultaneously, however, Du Toit noted that there were limitations to the study of glaciation as much of the evidence either lay buried deep beneath succeeding layers of rock or had been exposed, only to be completely eroded. The glacial deposits, therefore, which were visible and thus able to be studied, occupied only a small area, despite the vast extent of glaciation during the Carboniferous Period. Moreover, Du Toit also highlighted the complexity of interpreting glacial movement. Striations, for instance, only show the direction of movement at one point in time and do not indicate if the direction of flow subsequently changed. The presence of more exotic or unique rocks with their origins in particular formations in the glacial tillite allows the geologist to broadly trace the terrain over which the glacier passed but, if the tillite abounds with the more commonplace rock such as run-of-the-mill granite, it becomes more difficult.[12]

In this paper Du Toit challenged the existing view of a single ice sheet covering southern Africa, suggesting that the evidence indicated four

[11] Alex L. Du Toit, "The Carboniferous Glaciation of South Africa," in *Transactions of the Geological Society of South Africa* (Johannesburg: Geological Society of South Africa, 1921), p. 188.

[12] Du Toit, "The Carboniferous Glaciation of South Africa," p. 189. It should, however, be noted here that Du Toit was writing at a time when the technology for the detailed analysis of rock chemistry was still in its infancy.

separate ice sheets—the two most prominent were the Transvaal and Natal glaciers and the lesser glaciers were the Griqualand West and Namaland Ice (the evidence for the latter, however, was extremely meagre, and its status could be reconsidered if further evidence was unearthed). Particular attention is paid to the Transvaal glacier—although much of its depositional history was either buried beneath younger strata or was completely eroded. When glaciation was at its height, the Griqualand glacier was subsequently incorporated into the Transvaal glacier.[13] The Natal glacier appeared to have begun in the Indian Ocean and subsequently displaced the older Nama glacier, which left trace evidence in parts of the Dwyka Formation. The tillite of the Natal glacier was akin to that found in southern Madagascar and, for Du Toit, this was evidence that the glacier had covered both regions, implying that they were in closer proximity in the past. The four glaciers that Du Toit was able to identify then joined to form a single massive ice sheet that covered southern Africa, moving south until it reached water. Its northern terminus was less clearly defined and the glacier was less substantial here due to it being closer to the equator and, hence, warmer.[14]

Towards the conclusion, Du Toit portrayed a world that was completely alien to the contemporary reader where "The Ice formed a continuous sheet, probably thousands of feet in thickness, burying the topography completely, though the tops of the highest ranges … may have projected". This ice sheet terminated in fresh water rather than ocean, an indication "that a *second barrier either of land or ice must have lain further to the south*" [emphasis in original].[15]

While the origins of the glaciation remain shrouded, for Du Toit, it was clear that it could not be accounted for if the landmasses were in the same position that they currently occupy. What the pattern of glaciation indicated was that it had covered a single continent that had thereafter fragmented. This was the hypothesis that best fit the visible evidence. The glaciation then became the means by which Du Toit was able to surmise the appearance of Gondwana during the Carboniferous and Permian Periods. In the paper, this is depicted in the form of three rough sketches

[13] Du Toit, "The Carboniferous Glaciation of South Africa," pp. 193, 196.

[14] Du Toit, "The Carboniferous Glaciation of South Africa," pp. 191, 193, 196, 200–201, 203, 207.

[15] Du Toit, "The Carboniferous Glaciation of South Africa," p. 213.

based on the glacial evidence, showing a single landmass partly lying on the South Pole and reaching the 45th parallel.[16]

In Du Toit's use of glaciation to underpin the theory of continental drift, he was hampered—like Wegener and Suess—by the lack of understanding regarding the process. Early work on glaciation—drawn upon in Du Toit's work—was based on simple models. Louis Agassiz, the prominent Swiss geologist had determined periods of glaciation based on their tillite remnants. The model proposed by Albrecht Penck and Eduard Brückner in 1909 consisted of four periods of Ice Age glaciations that was still used 40 years later. It was only with the development of radioactive dating in the mid-twentieth century and the use of deep sea and ice cores (drawn largely from Antarctica) that more accurate and detailed representations of climate became possible. The analysis of oxygen isotopes in *foraminifera* obtained from sea cores reveals that a greater proportion of the heavier ^{18}O (consisting of eight neutrons and six protons in the nucleus, with the mass of an atom derived from its nucleus) relative to the common ^{16}O (consisting of six neutrons and six protons in the nucleus) isotope in the oceans is indicative of glaciation. Core sample analysis also vindicated the work of Milutin Milankovitch, an engineer who, beginning in 1912, worked for three decades on developing a mathematical hypothesis showing the cyclical nature of the Earth's climate. Termed the Milankovitch cycles, he proposed that the advance and retreat of the polar ice caps was based on deviations in the form of the earth's orbit as well as variations in the earth's distance from the sun. The convergence of these factors had the potential to create either phases of glaciation or interglacial periods. These cycles occurred on vast timescales that contained a certain mathematical symmetry—every 21,000, 42,000 and 95,000 years. The reconstruction of past climate is a complex process, drawing upon a variety of different methods such as the study of plant remains and pollen as well as animal and insect remains, offering an indirect view of climate through the life that abounded during particular time periods, rendered increasingly accurate by radiocarbon dating.[17] While this technology and the use of absolute rather than relative dating was not available to Du Toit, he was able to use the geological and fossil evidence to argue for similar periods of glaciation, indicative of global changes of climate.

[16] Du Toit, "The Carboniferous Glaciation of South Africa," pp. 219, 220, 227.

[17] Brian Fagan, *Cro-Magnon: How the Ice Age gave Birth to the First Modern Humans* (New York, Bloomsbury Press, 2010), pp. 30–31, 28, 54–55.

Despite the radical interpretation of the past evident in "The Carboniferous Glaciation of South Africa" which would showcase Du Toit's increasing prominence as a drift theorist, it apparently made little impact on his South African contemporaries. An article appeared in the *Cape Times* in September 1922 that reported on a recent meeting of the Royal Society of South Africa where continental drift was the main topic of discussion. With no mention of the work of Du Toit, the meeting summarised the existing work by Suess, Taylor and Wegener on continental drift and its application to the southern landmasses. South African geologists were clearly divided. Professor A. Young's dissent took on a personal note as he highlighted Wegener's shortcomings, the main one being his lack of geological background and Young confidently asserted that "he knew of no responsible geologist who had accepted Wegener's views".[18] For Young, the similarities between South America and South Africa—that Du Toit would soon begin investigating—could be just as easily be explained by other means. He was countered by Sidney Haughton—author of the *Geological History of Southern Africa*—zoologist, HB Fantham of Wits and botanist Robert Compton who spoke of the similarities across continents between fossils, parasites and flora respectively. The final word, however, was the prerogative of the president of the Society, ichthyologist J.D.F. Gilchrist who silenced the prospective drift theorists with a somewhat trite, "similar conditions would produce similar forms, and that was a more likely explanation of any similarities". This served to end the meeting of the Society on an oppositional note.[19] Two decades later, Haughton would be more circumspect in his assessment of the value of faunal evidence pointing out that, while there were indeed faunal similarities across the various continents, there was as yet no common agreement on what constituted distinct Permian species and the fossil record was itself incomplete which promoted uncertainty. While it was possible that faunal similarities could lend weight to existing argument, Haughton concluded that faunal analysis on its own was insufficient to prove the interconnected nature of the southern landmasses.[20]

[18] UCT-JL: Alex L. Du Toit Papers—BC 722: B3 Correspondence: B4 Correspondence (Chapter XIV—Biological and Palaeontological). "Do Continents Float? Wegener's Theory Applied to Africa," in the *Cape Times*, 30 September 1922.

[19] "Do Continents Float? Wegener's Theory Applied to Africa."

[20] UCT-JL: Alex L. Du Toit Papers—BC 722: M—Folders: M8—Correspondence with Geological Survey and individual geologists, 1942–1947. Letter from SH Haughton, Director, Geological Survey, Pretoria to Du Toit, 13 May 1943.

The equivocation and hostility of his South African colleagues, however, did little to deter Du Toit. In this short period, it was clear that Alex Du Toit was a firm proponent of continental drift. His research, thus far, had focused on South Africa although he was able to draw on a wealth of evidence from other Gondwana landmasses that bolstered his own thinking. He was now poised to engage in research further afield.

CHAPTER 7

And Yet It Moves...: Du Toit's South American Journey

In the introduction to *A Geological Comparison of South Africa with South America*, Du Toit begins by confronting the prevailing theory of "land bridges" that had been used to explain the similarity of fauna and flora on continents separated by vast oceans. While this is not disputed in some cases such as, for instance, the Bering Strait that allowed the migration of humans and flora from the Old World to the New when sea levels were low, there was a tendency to assign any fossil or floral similarity to hypothesised land bridges with little substantiating evidence. For Du Toit, his challenge to the "land bridge" explanation takes on three forms: the first is the similarity of land-based rather than marine-based fossils; secondly, across the southern continents there exists a markedly similar stratigraphy suggesting a similar climate which, at the outset, was marked by ice sheets but became increasingly dry, culminating with the formation of desert environments. His final point here was that, based on an analysis of marine fossils found on land, the oceans themselves were relatively young.[1]

The title of this chapter is based on the English translation of the quote that has been credited to Galileo Galilei, indicating his defiance of the Inquisition that forced him repudiate his view that the earth was not the centre of the universe but revolved around the sun.

[1] Du Toit, *A Geological Comparison of South America with South Africa* (Washington, Carnegie Institution of Washington, 1927), p. 1.

S. Chetty, *'Africa Forms the Key'*, Cambridge Imperial and Post-Colonial Studies, https://doi.org/10.1007/978-3-030-52711-2_7

At the outset, then, Du Toit sets up his own work in opposition to prevailing doctrine. He then aligns his own observations within the findings of the prominent contemporary "drift" theorists—Eduard Suess, Frank Bursley Taylor and Alfred Wegener. Interestingly, however, Du Toit makes the claim that he had arrived at his conclusions independently from that of Wegener based on his own knowledge of the similarity between the "glacial deposits" of Australia and the Karoo. Armed with this understanding, he applied successfully to the Carnegie Institution, obtaining a research grant in order to further his research in South America.[2]

The grant from the Carnegie Institution created some intellectual strain for Du Toit's work in South America. While the reception of continental drift was largely sceptical for the first half of the twentieth century, a particular virulent hostility was present on the part of American geologists. The American response may have been due to their uneasy relationship with Europe and an assertion of American nationhood. As a result of the Scientific Revolution and strengthened by the Enlightenment, the modern scientific method is composed of induction and deduction. The former refers to an assemblage of facts based on observation, from which an explanatory framework or theory may emerge. Deduction, on the other hand, takes theory as a starting point and observations are then assembled to substantiate the postulation. Prior to modern geology, deduction was the dominant form where geologists developed theories explaining how the earth worked. However the work of Hutton and Lyell privileged observation and the accumulation of facts and, for Lyell in particular, it was the everyday observable geological processes such as erosion and deposition which underpinned uniformitarianism.[3]

Oreskes addresses the ways in which the various methodologies employed by scientists—and dependent upon their context in space and time—play an important role in determining how they interpret the world. It is the reason that the great majority of American geologists were antagonistic—and more so than geologists from other countries—to continental drift. The convention explanation for their rejection of drift theory was that the means used to explain movement were inadequate and Oreskes argues that this was not the case. The explanations put forward for the "mechanism" of drift were not beyond the realm of possibility. Nor was it

[2] Du Toit, *A Geological Comparison of South America with South Africa*, p. 2.

[3] Ted Nield, *Supercontinent: Ten Billion Years in the Life of Our Planet* (London, Granta Books, 2008), pp. 130–133.

due to Wegener being an outsider which is not an explanation that can apply to Du Toit. It instead challenged the methodologies employed by American geologists and was rejected on these grounds.[4] At the start of the twentieth century, American and European geologists held contradictory views regarding the changing nature of the earth. The latter were more likely to acknowledge drastic and ongoing change that transformed land into sea and vice versa. For the former, however, "Continents were always continents, oceans were always oceans, and change was confined to the discrete zones at the interface between them".[5]

Science became a means of asserting a particularly American identity and a new system was developed termed the "multiple working hypotheses". In this process facts were assembled and then various existing theories were tested for the "best fit". Once a suitable explanatory framework was identified it was further tested and polished by accumulating additional observations and information in a progression towards truth. Wegner's writing on continental drift almost seemed deliberately designed to rub American scientists the wrong way. At the outset, Wegener begins by highlighting the way in which he was taken by the fit between continents while perusing a map of the world which added a distinctly happenstance tone to his theory. With theory in mind, Wegener then apparently proceeds to assemble the facts to fit it, rejecting all other possibilities. In subsequent editions of his work, he also tended to adopt a more inflexible tone in response to criticism. Finally, he committed the cardinal sin by being a meteorologist and thus an "outsider" (despite his expertise in geophysics). This, then, went a long way in influencing American rejection of continental drift.[6] In addition to the "multiple working hypothesis", uniformitarianism served to confirm geology as a science, distinguishing it from Biblical catastrophism—and the break-up and movement of continents was perceived to be catastrophic. Moreover, uniformitarianism was predicated on observable processes whereas drift was inferred. On a technical note, Americans subscribed to the model of isostasy proposed by John Pratt which differed from that of Briton, George Airy. According to the latter, varied thickness in the earth's crust would lead to a corresponding variation in isostatic compensation, that is, the higher the mountain,

[4] Naomi Oreskes, *The Rejection of Continental Drift: Theory and Method in American Earth Science* (New York, Oxford University Press, 1999), pp. 5–6.

[5] Oreskes, *The Rejection of Continental Drift*, p. 19

[6] Nield, *Supercontinent*, pp. 132–135.

the deeper its roots. This depth would decrease as the mountain lost height due to erosion. For Pratt, on the other hand, the isostatic compensation would remain constant regardless of the density of the crust. For American geologists, Pratt's model seemed to bear out their own observations. In contrast, continental drift would result in increasing density of crust, particularly at convergence sites, which seemed to bear out the Airy model.[7]

For American geologists, Wegener and his predecessor, Suess, represented everything that was wrong with regards to their use of the scientific method—their conclusions were based on theory rather than the empiricism of fieldwork. They were dismissed as "armchair geologists".[8] As geologist John Dewey recollected when introduced to Tuzo's fault model in 1964, the "classically trained geologist [was] one [who] must always first gather lots of data, then build slowly upward towards models". The notion of beginning with the model was unfamiliar and, decades earlier, anathema.[9] "Armchair geologist", however, was certainly not an appellation that could be applied to Du Toit who had a history of undertaking extensive fieldwork.[10]

In 1922, Professor Reginald Daly, a geologist at Harvard University, spent nine months in Africa, paying particular attention to the Bushveld Igneous Complex in South Africa.[11] Reginald Daly—like Frank Bursley Taylor—was part of that unusual breed of American geologist who was a "mobilist", rejecting the permanence of oceans and continents espoused by his countrymen. While initially studying mathematics, Daly developed a fascination with geology, culminating in his appointment as Professor of Geology. Daly's focus had been on volcanism and igneous rocks and, like Du Toit, engaged in extensive fieldwork on volcanic islands such as the Hawaiian chain and St Helena. Through his research Daly came to believe

[7] Naomi Oreskes, "From Continental Drift to Plate Tectonics" in *Plate Tectonics: An Insider's History of the Modern Theory of the Earth*. Naomi Oreskes (ed) (Boulder, Westview Press, 2003), pp. 10–11.

[8] Nield, *Supercontinent*, p. 143.

[9] John F. Dewey, "Plate Tectonics and Geology, 1965 to Today" in Plate Tectonics: An Insider's History of the Modern Theory of the Earth. Naomi Oreskes (ed) (Boulder, Westview Press, 2003), p. 230.

[10] In something of an ironic twist decades after plate tectonics became the key explanatory framework for geology, some geologists were wary of the over-reliance on computer modelling and its prioritisation over fieldwork. John McPhee, *Annals of the Former World* (New York, Farrar, Straus and Giroux, 2000), pp. 148–149.

[11] Nield, *Supercontinent*, pp. 143–144.

that as a result of the pressure and temperature at depth, the rock that underlay the Earth's crust was molten. The high pressure meant that this molten rock did not behave like a liquid but instead like glass and, as such, formed the layer upon which the crust above it could move. Like Du Toit, too, Daly's hypothesis elicited no small amount of criticism with A.P. Coleman—the man who moved Daly to become a geologist—describing it as an act of "poetic imagination". The criticism evoked meant that Daly became more cautious in his advocacy of drift, a year later delivering "A Critical Review of the Taylor-Wegener Hypothesis" that was couched within a more appropriate level of scepticism for an audience of American geologists.[12] Not all receptions, however, were as hostile and in South Africa he was to find a kindred spirit.

It was during his visit to South Africa that Daly first met Alex Du Toit. Accompanying Daly on the expedition was Frederick Wright, a geologist attached to the Carnegie Institution which funded Daly's research in Africa. Upon his return to the United States—and influenced by Daly who had come to espouse drift theory—Wright attempted to persuade John Merriman, the Director of the Carnegie Institute of Washington D.C., to provide funding to undertake research into drift theory by examining geological evidence in South Africa and South America. This area of focus was based on a proposal made at a meeting of the British Association for the Advancement of Science. For Wright, Alex Du Toit was the ideal geologist to undertake the necessary fieldwork.[13]

Merriman initially refused Wright's request—and it had to be rephrased in light of the American scientific method. Du Toit was already a proponent of continental drift but his research could not be seen as a means of vindicating Wegener's theory as that was deductive reasoning. He was to apply instead the largely inductive multiple working hypotheses: "The stated main purpose of the trip would be to gather facts: to add these good, solid bricks to the great edifice of science, and then to offer many different explanations of them, drift being merely one theory among equals".[14] It is thus with this contradiction between his own beliefs and that of the organisation that funded him, that Du Toit left for South America.

[12] Roger Muir Wood, *The Dark Side of the Earth* (London, George Allen and Unwin Publishers Ltd, 1985), pp. 80–81.

[13] Nield, *Supercontinent*, pp. 143–144.

[14] Nield, *Supercontinent*, p. 145.

The result of his journey was *A Geological Comparison of South America with South Africa*, published by the Carnegie Institution in 1927. Du Toit emphasised two key aims in writing the book. The first is the marshalling of evidence from the east coast of South America, the west coast of Africa and the Falkland Islands in support of "displacement theory". He anticipated the antagonistic reception to continental drift and his book therefore assembles a wealth of evidence in order to persuade the reader:

> The schematic representation embodied in Figure 7[15] will assuredly be subjected to active and probably to hostile criticism on the score of the fantastic and apparently improbable character of the displacement theory, but a close and impartial study thereof is invited.[16]

Reasoned and impartial criticism is welcomed as is the postulation of alternate valid hypotheses.[17]

Du Toit's secondary aim was to provide an account of South American geology and he describes his work as "the first comprehensive summary of the geology of the eastern part of South America". Simultaneously, however, he highlights the difficulties of so ambitious a project in terms of the vast distances to be explored and geologically mapped, the lack of a pooling of knowledge between these states that allow for a complete understanding of the region's geology and the time constraints to which he was subject. Du Toit was thus compelled to rely in some instances on the work of geologists on the continent—one of whom was geologist, Juan Keidel, who Du Toit cited extensively.[18]

In the first half of the twentieth century, most of the geologists working in Argentina had emigrated from Europe and, in particular, Germany. One of whom was Gustavo A. Fester who taught geology at the *Instituto Nacional del Profesorado* in Parana, was involved in studying the Andes Mountains and formed part of a network of Argentinian-German geologists that included Juan Keidel, a particularly noteworthy figure. Keidel's pioneering work in the Sierra de la Ventana mountain range in the province of Buenos Aires was heavily drawn upon by Alex Du Toit and

[15] A visual depiction of the "fit" between the coastlines of South America and Africa based on their stratigraphic correlation. Cf Du Toit, *A Geological Comparison of South America with South Africa*, p. 116.

[16] Du Toit, *A Geological Comparison of South America with South Africa*, p. 4.

[17] Du Toit, *A Geological Comparison of South America with South Africa*, p. 4.

[18] Du Toit, *A Geological Comparison of South America with South Africa*, p. 3.

A Geological Comparison of South America with South Africa demonstrates the extent of Keidel's influence, with particular reference to the Gondwanides. The term had been coined by Keidel and referred to a Gondwana-era mountain range that included both the Sierra de la Ventana and the Cape Fold Belt of South Africa. Initially part of the same mountain range due to the trend of their folding and their stratigraphy, the two were now separated by an ocean. The overwhelming evidence derived from the southern landmasses (and India) that comprised Gondwana, along with the evidence provided by figures like Juan Keidel meant that geologists in the southern hemisphere were more favourably disposed towards continental drift theory than their northern counterparts. A female student of Fester was little surprised when plate tectonics emerged as a viable theory decades later as she had already been exposed to drift theory in 1939.[19]

Du Toit sailed for South America from Cape Town on the 12th of June, 1923, arriving in Rio de Janeiro two weeks later. Despite his very detailed diary entries for other years, it is of some consternation to researchers that the diary entries for this seminal year end with the characteristically brief:

> "Thursday, 31st May—"Packing and getting ready"
> Wednesday, 6th June—"Left in evening by 7.20 for Cape Town and S. America"[20]

Rather endearingly, subsequent entries for that year penned in a different hand are clearly written by Du Toit's son, Robert, relating his own experiences including numerous letters to his father during the latter's five-month absence.

Du Toit's fieldwork in South America is recorded in the form of a notebook that follows the format of his earlier trip to Australia—containing detailed descriptions of the geology accompanied by sketches of fossilised flora, all of which would form the content for *A Geological Comparison of South America with South Africa*. Also mentioned in his notes are earlier surveys and conclusions drawn by Juan Keidel, indicating his reliance on work already done in South America. During the course of his travels

[19] Carlos W. Rapela and Pedro J. Depetris, "Geochemistry in Argentina: from pioneers to the present" in *Environmental Earth Sciences,* 75:524 (2016), pp. 7–9.

[20] UCT-JL: Alex L. Du Toit Papers—BC 722: L: Alex Du Toit material transferred from Geological Sciences, Oct 2005: Field notebooks.

through South America he received much assistance from geologists and their associated institutions who are thanked extensively in the opening chapter of his monograph.[21]

Examining a rock formation in Ollavarria, Argentina (University of Cape Town Libraries, mss_bc722_s_south_america_101a)

The remainder of the work focused on a detailed description of the geology of South America and its correlation with parts of South Africa, India and Australia—the territories that comprised the large southern continent of Gondwana. The second chapter therefore begins with a detailed discussion of the geology with which Du Toit was most familiar—that of South Africa. He proceeds further north along the east coast until the region of the Congo, all the while highlighting the points of similarity based on rock formations and composition, stratigraphy and fossil evidence. Of particular importance to Du Toit here was the Falkland Islands which he perceived to be the "most striking link between South Africa and South America". Some areas of similarity included those between Falklands sandstone formations and that of Table Mountain, the pervasive presence of the *Glossopteris* fossil flora and the correlation between the folding

[21] Du Toit, *A Geological Comparison of South America with South Africa*, pp. 4–5.

patterns and age of mountains on the islands with that of the Western Cape. This suggested that, not only were the Falkland Islands representative of an "intermediate" stage of continuity between South Africa and Argentina, they were actually closer geologically to South Africa than to South America despite their closer proximity to the latter.[22]

In his discussion on Argentina, Du Toit makes use of the earlier work of Juan Keidel. While Du Toit relies extensively on the earlier mapping and exploration of Keidel he does, at times, disagree with the Argentinian geologist's conclusions. This is particularly apparent in the interpretation of the fossil record where Keidel suggested that there were two clear fossil layers—one dating from the Gondwana period and another earlier layer belonging to the Carboniferous Period which ended approximately 299 million years ago and was marked by flora originating in the northern latitudes. This, of course, adds a level of complication to Du Toit's argument which focuses on Gondwana and the interconnectedness of largely southern land masses. While acknowledging that further research is needed, he still makes a substantial attempt to argue that his observations do not suggest two distinct fossil layers and that the overwhelming presence of *Glossopteris* and similar fossil evidence favours only Gondwana. At the same time, *Glossopteris* was an enduring plant and could be traced as far back as the Carboniferous Period.[23]

Despite his use of Keidel's data, Du Toit's feelings towards Juan Keidel were somewhat ambivalent, demonstrating that concordance on drift did not necessarily translate to intellectual harmony. Du Toit's necessarily more cursory analysis of the significance of South American geology had been built on the work of Keidel yet the two men were not in agreement over the interpretation of the geological strata with Du Toit believing that Keidel "appeared to have developed a distinctly antagonistic attitude towards my few discoveries, both in the Pre-Cordillera and Sierras de B.A."[24] The use of "antagonistic" suggests a serious difference of opinion. This is presaged in earlier correspondence where Du Toit displays what can only be termed a sense of irritation towards Keidel who had played a

[22] Du Toit, *A Geological Comparison of South America with South Africa*, pp. 8–9, 11, 12–13.

[23] Du Toit, *A Geological Comparison of South America with South Africa*, pp. 37–39.

[24] UCT-JL: Alex L. Du Toit Papers—BC 722: B1—General Correspondence (1930s–1940s): Letter by Du Toit to Dr Arnold Heim, Buenos Aires, 9 October 1945.

significant role in mapping South American geology, alongside Horacio Harrington.

Horacio J. Harrington was a South American geologist who had conducted fieldwork research alongside Keidel. Born in Argentina in 1910, Harrington received his doctorate in the natural sciences from the University of Buenos Aires and took part in the geological survey of Staten Island in South America. He was subsequently awarded a scholarship to study at Oxford University where he was awarded a PhD in Geology. Practising as a geologist at *Dirección Nacional de Minas y Geología* and holding teaching positions at the University of Argentina and University of La Plata, Harrington was an acknowledged expert on fossil evidence and the geology of Argentina and other parts of South America.[25] While Du Toit acknowledges the valuable input offered by the joint fieldwork of Harrington and Keidel regarding glaciation and the need for supporting fossil evidence to correlate periods of glaciation, he is nonetheless hostile to perceived criticisms of his own work by Keidel. Du Toit points out that his very brief time in South America meant that his knowledge was cursory and thus not deserving of Keidel's nit-picking:

> I am somewhat annoyed at the tone of Keidel's long paper in the Geol. Rundschau of Jan. Instead of going right ahead & giving in full detail all his own observations & mentioning the chief points of difference from myself, he analyses every statement of mine meticulously, as though my work pretended to be a final picture of Argentine geology. So on every page of his paper appears 'Du Toit this' 'Du Toit that', ad nauseam. Considering that all the time I had to spend on the two regions in question was about 10 days, that my guides & colleagues from the Geological Survey did not know the areas at all, I think I did quite a lot, but I certainly never doubted that detailed work would modify even wipe out much of my tentative conclusions.[26]

There does appear to be, therefore, a clear sense of rivalry between the two men with Du Toit's work in South America possibly viewed as trespassing on Keidel's area of expertise and Du Toit's subsequent

[25] John J. De Benedetti, "Memorial to Horacio J. Harrington, 1910–1973", http://rock.geosociety.org/pub/Memorials/v07/Harrington-HJ.pdf, Accessed 1 September 2017.

[26] UCT-JL: Alex L. Du Toit Papers—BC 722: B1—General Correspondence (1930s–1940s): B3.2 Correspondence: D-H: Letter by Du Toit to Horatio J. Harrington, 14 March 1939.

defensiveness regarding the limited time he had available to come to grips with South American geology.

Despite his disagreement with Keidel, Du Toit would acknowledge the necessary limitations of his observations made in South America and turn to South American—as well as Indian and Australian—experts for greater clarity as he reconstructed Gondwana and retained a sense of hesitancy in the claims he was able to make for the geology outside of South Africa: "It must candidly be admitted that today the stratigraphy of the Gondwana in your country is somewhat confused & I feel rather reluctant to add thereto by further comments, realising that my journeys in your most interesting land have been so brief & geological work so restricted".[27]

The final chapters of Du Toit's monograph assemble the primary and secondary evidence obtained by Du Toit in order to refine, not only Keidel's findings, but that of Alfred Wegener as well. Rather than seeing "the New World ... as having parted from the Old", Du Toit suggests instead that there were two original continents—one in the northern hemisphere that gave rise to North America and Eurasia and Gondwanaland in the south that comprised the present landmasses of South Africa, South America, India and Australia. The contact that these two continents may have had with each other led to the formation of mountain chains such as the Atlas Mountains in North Africa and the Alps in Europe.[28]

In his work at this point, Du Toit had considered the significance of the Mid-Atlantic Ridge but he, like Taylor and Wegener before him, had not arrived at an adequate explanation for the movement of the continents. It was instead geologist Arthur Holmes who in 1929 had considered the heat released by the decay of radioactive elements within rock. Holmes believed that this heat played a role in the movement of continents by creating convection currents in the mantle. He used the Scotia Arc to demonstrate this, suggesting that it was a result of the movement of the South American and Antarctic "continental blocks".[29] And, as with most aspects related to continental drift, he was not the first to do so. The

[27] UCT-JL: Alex L. Du Toit Papers—BC 722: B2—Correspondence with colleagues in South America (ca. 1920s–1931): Letter by Du Toit to Sr E. Terra Arocena Ingeniero-Director, Instituto de Geologia y Perforaciones, Montevideo, Uruguay, 1 November 1928.

[28] Du Toit, *A Geological Comparison of South America with South Africa*, pp. 119–120.

[29] Philip Stone "Geological exploration of South Atlantic islands and its contributions to the continental drift debate of the early 20th century" in *Proceedings of the Geologists' Association,* 2015, Vol 126, 266–281, p. 10. Taken from http://nora.nerc.ac.uk/510724/1/SouthAtlanticIslands-pga.pdf, Accessed 12 October 2016.

notion of convection as providing a mechanism for movement had been bandied about since as early as 1839. The metaphor used was that of pea soup atop a heated surface. The heated liquid would rise, displacing surface material before cooling, moving back down, only to rise again. *The Physics of the Earth's Crust* penned by the Reverend Osmond Fisher had the distinction of being the first book explaining geology through the lens of geophysics and had also considered convection as responsible for geological features. In the decade following Holmes' proposition, actual experiments were carried out that employed "rotating cylinders in tanks full of goo covered with thin skins of wax" that gave some inkling as to how convection currents could be responsible for continental drift.[30]

The Evidence: Fossils and Formations

In *A Geological Comparison of South America with South Africa* a considerable amount of evidence is presented to the reader that can be almost overwhelming—it is clear that the sheer weight of information—which draws upon the fossil and stratigraphic record—is designed to present an unassailable argument for the southern continent of Gondwana.

Much of the evidence obtained by Du Toit comes from the Karoo Supergroup. This geological formation is roughly 12 kilometres thick and is vast—around 700,000 square kilometres. This Supergroup is largely defined by its sedimentary deposits containing a wealth of fossil material. Also evident here is material deposited as a result of glacial activity as well as igneous deposits associated with volcanic activity. With its origins in the late Carboniferous Period, the Karoo Supergroup also contains large coal deposits that have been exploited. Its unique stratigraphy and wealth of fossil evidence therefore made the Karoo Supergroup ideal for Du Toit's hypothesis. He was able to demonstrate that many of its defining features were shared by formations in the Gondwana landmasses, indicating their shared origin. Fossils provided a significant means of correlation.[31]

The importance accorded to the fossil record is Du Toit's inclusion of an appendix "Upper Carboniferous Fossils from Argentina" written by

[30] Nield, *Supercontinent*, p. 147.

[31] M.R. Johnson, C.J. van Vuuren, J.N.J. Visser, D.I. Cole, H. de V. Wickens, A.D.M. Christie, D.L. Roberts and G. Brandl "Sedimentary Rocks of the Karoo Supergroup" in *The Geology of South Africa*, M.R. Johnson, C.R. Anhaeusser, R.J. Thomas (ed) (Johannesburg and Pretoria, The Geological Society of South Africa and the Council for Geoscience, 2006), p. 461.

F.R. Cowper Reed of Sedgwick Museum, England, which includes detailed descriptions of the fossil samples provided by Du Toit. Du Toit himself concludes the monograph with a strong recommendation for a multidisciplinary approach on the part of "geologists, palaeontologists, zoologists, botanists and physicists" with a particular emphasis on South America and South Africa where the evidence for "displacement" is the strongest.[32] In Du Toit's hypothesis Gondwana, then, becomes more than an offshoot of the "Old World".

An important example of fossilised fauna cited by Du Toit is that of *Mesosaurus*. This was a reptile of approximately 70 centimetres in length that lived in water. Its fossilised remains were found in the Whitehill Formation that occurs within the Ecca Group.[33] Like the Dwyka Group, the Ecca Group forms part of the Karoo Supergroup and contains a total of 16 formations, the largest of which are the Prince Albert Formation and the Whitehill Formation.[34] There is a clear correlation in terms of fossil evidence as well as rock composition that suggests that the Whitehill Formation corresponds to the Irati Formation that is found in South America. The Whitehill Formation is a layer of mudstone that, along with the Irati layer were formed in a sea that had once existed between South America and Africa. Mudstone is formed as a result of sediment deposited in shallow coastal areas. With regards to the Whitehill Formation, while it is black in colour, rock exposed to weathering turns white, making it a distinctive and easily observable rock layer. *Mesosaurus* apparently inhabited this shallow sea and the numerous occurrences of its fossilised remains indicate the closer proximity of the two continents.[35]

[32] Du Toit, *A Geological Comparison of South America with South Africa*, pp. 129–150, 121.

[33] Colin MacRae, *Life Etched in Stone: Fossils of South Africa* (Johannesburg, The Geological Society of South Africa, 1999), p. 168.

[34] Johnson et al. "Sedimentary Rocks of the Karoo Supergroup", p. 465.

[35] MacRae, *Life Etched in Stone*, pp. 170–171.

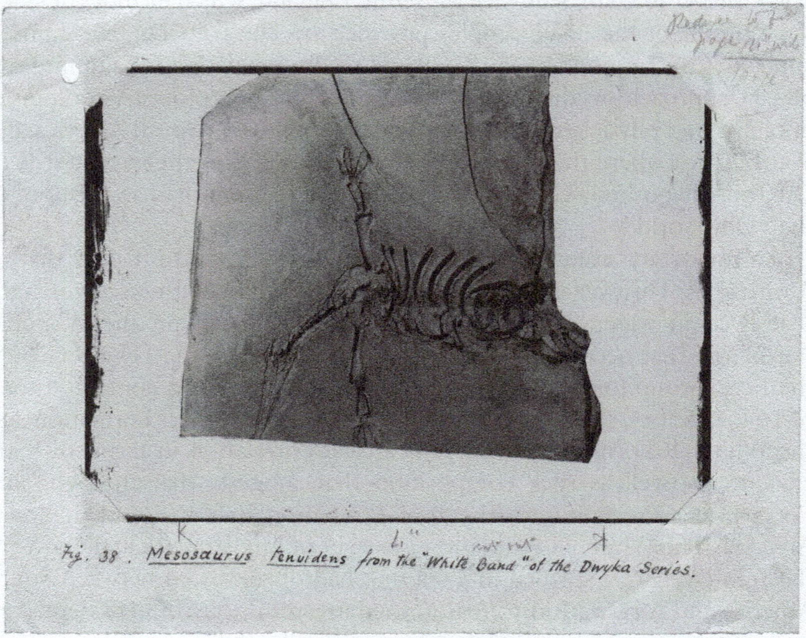

Fossilised *Mesosaurus tenuidens* (University of Cape Town Libraries, mss_bc 722_c2_4_002)

Even as Du Toit discussed fossil correlations in detail, he was aware that a significant criticism of the use of fossils as evidence was that they did not preclude "land bridges" as both plants and animals can migrate. They could therefore have appeared on currently distinct continents via connecting land bridges that had subsequently been submerged—as is the case with the Bering Strait. The same could not be said for rock strata and that is where Du Toit's evidence is most persuasive. Rock formations separated by continents sometimes demonstrated greater congruity than they did with formations on the same continent. For Du Toit there was only one interpretation—the similarity across rock strata suggested the proximity of separate continents in the distant past that were subject to identical geological and climatological processes.[36]

[36] Nield, *Supercontinent*, p. 146.

In addition to fossils, for Du Toit a visible indicator of the link between separate landmasses was the common deformation of rock, evident in the Cape Fold Belt. Here, over a period of time, large areas were lifted and folded in a process of orogeny. According to Du Toit, this Belt was part of a larger mountain system—the Gondwanides—and a similar folding pattern can be traced in the Falkland Islands, the Sierras Australes of Argentina and the Ellsworth Mountains of Antarctica, suggesting that all four were part of this single mountain chain. While more research is required regarding the various processes that led to the creation of the Cape Fold Belt, the direction of folding in conjunction with the evidence from the other Gondwana landmasses is an indicator of the ways in which these landmasses joined together as part of Gondwana.[37]

The publication of Du Toit's research and conclusions reached in South America in 1927 had a marked influence on pioneer drift theorists, not least of which was Alfred Wegener. The 1929 edition of *The Origin of Continents and Oceans* differed in one respect from the three earlier editions—it made use of the evidence obtained by Du Toit. Wegener paid particular attention to the stratigraphic correlations between South America and South Africa, highlighting not only the presence of diamond-bearing kimberlite pipes in Brazil and South Africa but also the tillite produced by glaciation during the Permo-Carboniferous Period and highlighted by Du Toit. Of particular relevance was the direction of glaciation as reconstructed by Du Toit through an analysis of striations and glacial deposits as well as the significance of finding similar glacial evidence across the Falkland Islands, in Brazil, Argentina, India and Australia. Wegener went on to write a detailed summary of the main points of evidence assembled by Du Toit to give credence to continental drift, assessing the South African's important contribution to drift theory: "The whole work is a unique geological demonstration of the correctness of drift theory so far as these parts of the globe are concerned".[38]

Wegener also acknowledged and incorporated within his own understanding the refinements Du Toit had made to continental drift. These included the importance given to the Falkland Islands and its closer

[37] A.R. Newton, R.W. Shone and P.W.K. Booth "The Cape Fold Belt" *The Geology of South Africa*, M.R. Johnson, C.R. Anhaeusser, R.J. Thomas (eds) (Johannesburg and Pretoria, The Geological Society of South Africa and the Council for Geoscience, 2006), pp. 521–523, 527–528.

[38] Alfred Wegener, *The Origins of Continents and Oceans*, John Biram (translator) (New York, Dover Publications Inc, 1966), pp. 64–65, 130–131, 68–69.

geological correlation to South Africa rather than South America as well as the hypothesised reconstruction of the proximity of South America to Africa including the suggested gap between the two of between four and eight hundred kilometres that allowed for a shallow sea, that created conditions for the deposition of fossils and would later be the site of rifting between the two continents. The preponderance of evidence as a result of the extensive fieldwork carried out by Du Toit made no small impact on Wegener and was met with resounding approval: "I must admit that Du Toit's book made an extraordinary impression on me, since up till then I had hardly dared to expect so close a geological correspondence between the two continents".[39] It was the scientific method exemplified—the convergence of the deductive and inductive methods and was also an indication of Alex Du Toit's prominence as a drift theorist. Despite this growing prominence the period also saw him continue in his role as a teacher and academic—where he was able to introduce a younger readership to his findings.

Du Toit's career began as a teacher of geology in Scotland and this was a role that he was to play for the rest of his life. His diary entries reveal that correspondence was a daily aspect of his routine as he addressed queries from geological enthusiasts and peers from all over the world. This would continue even after his retirement. The time allocated to this proved exasperating and Du Toit considered the "time-consuming drudgery of correspondence—private and public" a distraction from his research.[40] In the midst of his research on continental drift, Du Toit was involved in the more humdrum activities of academia—paperwork, examinations and lectures. Teaching was also a significant part of his legacy and is embodied in two publications.

The Geology of South Africa

The Geology of South Africa is a comprehensive survey of the country's geological history and stratigraphy. The first edition was completed in 1926 and Du Toit was at work on the third edition at the time of his death

[39] Wegener, *The Origins of Continents and Oceans*, pp. 71–72.

[40] UCT-JL: Alex L. Du Toit Papers—BC 722: A4—Personal Documents: rough biographical notes.

in 1948.[41] While much focus has been on Du Toit's research into continental drift, *The Geology of South Africa* is a significant work that captures all the geological knowledge of the country and reveals Du Toit's tremendous command of the discipline. Perceived to be a textbook, the opening chapter is a description of general geological processes such as erosion, deposition and the notion of geological time as determined by stratigraphy. It is even at this early stage that Du Toit advocates the need to see the "bigger picture"—geology was not simply a descriptive science confined to a particular region but entailed a much broader history of the Earth itself. To this end, Du Toit portrayed a holistic geology that concerned itself with the past and continuing processes that shaped the earth, the comparisons between continents, changing climate and the evolution of life.[42] Here, as evident throughout his career, there is a multidisciplinary inclination that would be so necessary to plate tectonics decades later.

Subsequent chapters focus on the geology of South Africa in particular, outlining the various strata, their composition, their locations, their economic importance and, where appropriate, their fossil content. Yet Du Toit could also not resist highlighting the similarities between South African strata and those further afield. In his chapter on the Cape System, he relates the glacial strata deposited during the Carboniferous Period to those of the Falklands, the Sierra de la Ventana in Argentina and Paraná in Brazil, concluding with:

> This wonderful lithological, palaeontological, and structural parallelism between South Africa and South America during this epoch cannot be sufficiently emphasised. Taken in conjunction with the evidence of a similar kind during the Permian and Triassic, it proves the intimate connection of these two continents over an enormous period of time.[43]

Students of geology would thus be introduced, from the outset, to continental drift and the hypothesised Gondwana.

This is emphasised in his subsequent chapter on the Karoo System within which is located the Dwyka Conglomerate from which much of his glacial evidence is drawn. Du Toit describes an extensive system that

[41] S.H. Haughton, "Alexander Logie Du Toit, 1878–1948" in *Obituary Notices of Fellows of the Royal Society*, Vol 6, No 18 (November 1949), pp. 387–388.

[42] Alex L. Du Toit, *The Geology of South Afri*ca (London and Edinburgh, Oliver and Boyd, 1926), p. 11.

[43] Du Toit, *The Geology of South Africa*, p. 199.

covered much of present-day South Africa and extended even further to Madagascar, South America, Australia, India and Antarctica. Through the examination of striations—the long grooves left on the bedrock by the passing glacier and the presence of erratics—rocks transported far from their point of origin by the ice, in this case almost 1300 kilometres—Du Toit reconstructed the glacial movements, hypothesising that there were several such movements indicative of "independent centres of glaciation". The Karoo System also contained fossilised remains of *Mesosaurus* as well as the mollusc *Eurydesma* in the Northern Cape and present-day Namibia which compared favourably with fossils found in India and Australia.[44]

Concomitant with his holistic view of geology Du Toit included a separate chapter on the flora and fauna of the Karoo System. A major portion of this covers *Glossopteris*, a southerly variant of which could be found at the end of the Carboniferous Period and is usually evident in stratigraphic layers related to glacial periods. *Glossopteris* is particularly evident in the Lower Beaufort Formation, along with other floral remains that are also evident in New South Wales and India. The relative paucity of floral evidence in Middle Beaufort—with the exception of *Glossopteris* suggested to Du Toit that the southern landmasses were cut off from their northern counterparts. This isolation was, however, incomplete and he hypothesised that *Glossopteris* and the reptilian fauna of the Karoo were able to migrate to Northern Russia via India—an indication of the almost surreal conception of space required for drift theory. Du Toit subsequently adds a wealth of stratigraphic and further fossil evidence to suggest that the western region of Gondwana eventually split into two—the landmasses now separated permanently by the Atlantic Ocean.[45] The sheer weight of evidence used as illustration appears designed to convince, merely by its preponderance. The subsequent shift in the discussion to once again describe South African geology is marked, lacking the sense of argument, the desire to persuade.

A chapter of *The Geology of South Africa* is devoted to "Primitive Man". Du Toit begins by describing the various stone tools that make up the archaeological record in South Africa, highlighting the fact that the preservation and stratigraphic position of these artefacts are different to that apparent in Europe yet there is a general trend in the evolution of artefacts that is akin to that found in the northern hemisphere. He describes in

[44] Du Toit, *The Geology of South Africa*, pp. 200, 212, 214.
[45] Du Toit, *The Geology of South Africa*, pp. 270–271, 272, 278.

detail the various types of stone artefacts found in South Africa and first identified in Europe.[46] This section also reflected Du Toit's own interest in archaeology evident in his position as president of the South African Archaeological Society.[47] It is, of course, not the stone tools but the hominin remains that had drawn the world's attention to the tip of Africa. *Homo rhodesiensis* had been found in a Rhodesian cave in 1921: a Boskop skull—which Robert Broom posited as being the predecessor of the San—was discovered near Potchefstroom. However, pride of place was given to Dart's *Australopithecus africanus*, at the time controversially hypothesised to be the "missing link"—the common ancestor of modern humans and their ape cousins. For Du Toit, this suggested the exciting possibility that southern Africa was the birthplace of modern humans, the site of their evolution. At the very least he draws parallels between early humankind in Europe and "Bushmen" in South Africa based on cultural similarities such as cave paintings, suggesting that the latter evolved from the former and inextricably linking Africa with Europe.[48] Du Toit's thinking possibly had its antecedents in the proposition put forward by French prehistorian, Abbé Henri Breuil, in a meeting of the South African and British Associations for the Advancement of Science in 1929. Breuil was energised by what he perceived to be similarities between San rock art and that recently discovered in Spain, hypothesising that humankind had originated in the Sahara and subsequently moved both north into Europe and south into southern Africa. San rock art was therefore seen as a continuation of a cultural tradition—that had long since disappeared from Europe.[49]

Du Toit incorporated the history of humankind within the larger geological history of South Africa in his concluding chapter. Due to its very antiquity this early history is an obscure one but Du Toit suggests that the origins of the continent lay as far back as a billion years ago with the eruption of igneous material and the deposition of sediments. It follows the conventions of uniformitarianism—a hallmark of geological thinking for much of the twentieth century—with the repeated laying down of strata, paralleled by similar processes in India, Australia and Brazil. The formations that he described in such detail earlier are related to their origins in

[46] Du Toit, *The Geology of South Africa*, pp. 373–376.

[47] Haughton, "Alexander Logie Du Toit, 1878–1948", p. 391.

[48] Du Toit, *The Geology of South Africa*, pp. 376–378.

[49] Saul Dubow, *Henri Breuil and the Imagination of Prehistory: 'Mixing up Rubble, Trouble and Stratification'* in South African Archaeological Society Goodwin Series 12: 2019, p. 4.

volcanism, submergence, sedimentation and glaciation where seas became deserts and plants turned to coal. Again, it is Du Toit's ability to evoke lost geological worlds arising from the observable strata that is most suggestive here. And his conclusion relates this ancient geology to ancient humanity, calling for greater co-operation between geology and anthropology to represent South Africa's importance for the history of the earth and the humans who dwell upon it.[50]

Physical Geography for South African Schools

While *The Geology of South Africa* was written for the geology student, *Physical Geography for South African Schools* was written for a younger readership. First written in 1912 and reprinted in 1926, the textbook was designed to give students a rudimentary knowledge of climate, geography and geology with a particular focus on South Africa. As Du Toit asserted in the preface, prevailing publications—produced in Europe and North America—tended to focus on the northern hemisphere and did not always readily lend themselves to conditions south of the equator. The textbook was also envisaged, then, as a means of addressing a knowledge gap in South Africa—the dearth of scientific publications dealing with local conditions.[51] From the preface then—which Du Toit penned in December of 1911—he was clearly a proponent of the production of a particularly South African form of scientific knowledge, arguing for the new nation's distinctive character.

Du Toit's view of education—and his emphasis on science—can be contrasted to the more limited view of education envisaged for Africans in particular and reflective of the divisions and inequalities in South Africa. Charles T. Loram was already a prominent figure in liberal thinking regarding race in South Africa. A graduate of the Cape of Good Hope with a PhD from Columbia University, Charles T. Loram had published *The Education of the South African Native* in 1917. In this book, Loram saw the solution to the "native question" (the most effective means by which the small white population could effectively govern the majority black population in the light of modernity) as lying in education and the creation of a single, coherent state policy to oversee "Natives". Loram was

[50] Du Toit, *The Geology of South Africa*, pp. 422–433, 444.

[51] Alex L. Du Toit, *Physical Geography for South African Schools* (Cambridge, Cambridge University Press, 1926), pp. v–vi.

particularly keen on "industrial education" which could inculcate the useful practical skills, so necessary for a modern state, into the black working and aspiring middle classes.[52] Du Toit and Loram's views reflected the divisions between the academic and the vocational, divisions that took on a racial cast.

Covering a formidable range of topics, geology nevertheless lay at the heart of Du Toit's work and he contextualised geography—concerned with the contemporary physical surface features of the earth—with geology which offered the means by which these features had come into being. His argument was imbued with uniformitarianism, the belief that present processes offered a key to understanding past geological processes. In this light, the first chapter entitled "The Earth" offers a tantalising hint at Du Toit's possible reference to drift theory yet could also relate to the less contested notion of submergence and elevation as providing an explanation for the changing surface of the earth: "There is abundant evidence also to prove that in past geological times the distribution of land and sea was very different from that of the present day".[53]

While demonstrating a more than passing familiarity with atmospheric conditions in relation to their impact on climate, the textbook emphasises geological features. The chapter "The Ocean" addresses both the uneven distribution of landmasses with most of the land in the northern hemisphere and a correspondingly larger proportion of ocean in the southern hemisphere. Du Toit highlights the growing knowledge of the mysterious depths through increasing exploration and developments in communication and technology. Underwater cables offered the opportunity to measure the varying depths of the ocean as well as the possible composition of the ocean floor. Marine biology—a burgeoning field—was making great strides in understanding the life forms that lived beneath the waves. And, adding to the corpus of knowledge, were the intrepid research and exploratory missions conducted in the depths and in the polar regions—a demonstration of Du Toit's early interest in the potential of polar exploration.[54]

The descriptions of ocean topography are remarkable for the way they would eventually correlate with plate tectonics. There is a description of the "Dolphin or Challenger Rise", a ridge in the Atlantic Ocean that runs

[52] Richard D. Heyman, "C.T. Loram: A South African Liberal in Race Relations" in *The International Journal of African Historical Studies*, Vol 5, No1 (1972), pp. 42, 48.

[53] Du Toit, *Physical Geography for South African Schools*, pp. xii, 6–7.

[54] Du Toit, *Physical Geography for South African Schools*, p. 49.

the length of the Atlantic between Africa and North and South America. Du Toit hypothesised that this ridge may reach Bouvet Island—a suggested area of interest in the unrealised South African Antarctic Expedition. In addition to the ridges were the depressions and Du Toit describes the "troughs" or trenches found in the Pacific such as that off the coast of Japan. They correlate to an extent to the ridges and the coasts along which they are found. This may be an early description of subduction zones where one tectonic plate is subducted beneath another.[55]

The depths provided further evidence for the changing nature of the earth's surface. After addressing the characteristics of the ocean floor in terms of its rock composition, Du Toit pointed out that similar rocks could be found on land, suggesting that these land surfaces had once been submerged at great depth.[56]

Du Toit subsequently devotes great attention to the points where geology and geography overlap, addressing the role of wind in the creation of features such as the formation of sand dunes and loess as well as eroding the rock surface, an action that was compounded by temperature extremes. Other forms of weathering included chemical weathering where minerals in rock reacted with the atmosphere and further mechanical erosion through actions of plants and animals. Related to chemical weathering was the action of groundwater—a particular area of Du Toit's expertise—where chemical weathering of areas rich in limestone and/or dolomite created fissures and caverns that allowed for the collection of groundwater. Impermeable layers of rock also allowed for the accumulation of groundwater which could then be accessed through boreholes. This groundwater, when heated underground, allowed for the percolation and accumulation of mineral deposits.[57] Du Toit's discussion makes frequent reference to conditions pertaining to South Africa, thus fulfilling his aim at the outset.

From his discussion of weathering, he considers the agents of erosion—the role of rivers in shaping the landscape through the creation of valleys, waterfalls and alluvial plains. It was, however, glaciation that had a particularly transformative effect on the landscape. This, of course, was the basis for his early ground-breaking paper on glaciation during the Carboniferous that brought to public awareness his early affinity for continental drift. Du

[55] Du Toit, *Physical Geography for South African Schools*, pp. 49–51.
[56] Du Toit, *Physical Geography for South African Schools*, pp. 66–67.
[57] Du Toit, *Physical Geography for South African Schools*, pp. 68–74, 76–81, 85.

Toit distinguishes between two different types of glaciers—the valley or alpine glaciers that originate in mountains and ice sheets which expand outwards from a central point of origin, beginning in an area of relatively high topography. Extremely cold conditions allow for the accumulation of ice which, based on periods of freezing and thawing, subsequently begin to flow until they reach the sea. Here, pieces of the edge of the glacier break off to form icebergs. These glaciers were more extensive in the past than they are in the present and were responsible for dramatically shaping the landscape through their role in erosion and later, deposition. Evidence of glaciation can be determined from the features left behind and it was this evidence that ultimately led to Du Toit's supposition that an extensive ice sheet had covered Gondwana, with its origins in South Africa, further suggestive of a colder and drier climate. Evidence of glaciation included U-shaped valleys in contrast to the V-shaped valleys formed by rivers, the transportation of large rocks called "erratics" over vast distances which were subsequently stranded after the retreat of the glacier and terminal moraines, the unsorted debris at the edge of the glacier that is left behind. The bedrock may also bear striations caused by the movement over the rock-bearing glacier and these indicate the direction of glacial movement.[58]

Much of the geological aspect of the textbook then stemmed from Du Toit's own research into continental drift and in the subsequent chapter he made the transition from fire to ice in his consideration of volcanism. Again, there is a sense of prescience on his part as he points to the ways in which volcanoes appear to be grouped in particular areas and these areas correlate remarkably with the "oceanic depressions" or trenches. Current understanding of subduction zones indicates their affinity with volcanism. The subduction of sea-water containing oceanic crust raises the melting point of the magma leading to a weakening of the crust at these areas and the creation of volcanic island chains. Without the use of the technology that would become available decades later, Du Toit's explanation was remarkably correct: "The linear grouping is in many instances so marked that there can hardly be any doubt that such vents are arranged upon, and owe their existence to, a line of weakness or of fracture in the earth's crust some distance below the surface". He also addresses submerged volcanoes and volcanic "fissures" such as that present in Iceland—a landmass that is a product of the Mid-Atlantic Ridge.[59]

[58] Du Toit, *Physical Geography for South African Schools*, pp. 86–113, 116, 118–119, 122.
[59] Du Toit, *Physical Geography for South African Schools*, pp. 137, 138, 142.

But just as Du Toit appears to be on the verge of understanding the process of plate tectonics, his discussion of "crustal movements" is a demonstration of explanations for seismicity that were dominant at the time. Du Toit addresses two kinds of crustal movements—the first are the sudden movements evident in earthquakes while the second tend to be more sedate and are described as "secular movements" yet Du Toit attributes both to the same underlying mechanisms—uplift and subsidence. Evidence for the former is easily obtainable—the existence of marine fossils on dry land. Evidence for subsidence, however, is more elusive as it is usually found underwater; it can however be determined in the evidence of submerged forests that lie off the coasts or the rare obtaining of terrestrial fossils from offshore sources.[60]

The explanations for subsidence and elevation were firmly in the realm of the hypothetical and included the reduction of sea levels by extensive glaciation, the contraction of the earth's crust due to the belief that the earth had been steadily cooling since its creation, the role of erosion which decreased weight concomitant with deposition which increased it and volcanism and the movement of lava with its associated effects in the earth's interior. Without the evidence that underpinned plate tectonics, it was clear that geological activity lacked a unifying explanatory theory. Thus, while Du Toit is able to see the correlation between earthquake and volcanic activity as well as mountain-building, seismicity is explained solely in terms of subsidence and upheaval. It is the result of the abrupt movements of the earth's crust as it adjusts to these processes.[61]

Just as volcanism was linked to "troughs", Du Toit was able to make similar links to the formation of mountains—which were also associated with earthquakes. Suggesting that the reconstruction of Gondwana lurked at the back of his mind, he showed the pattern of mountain formation that was largely confined to edges of continents—a pattern that was evident largely south of the equator. Their formation was considered to be either due to selective erosion or was a dramatic indication of uplift. Du Toit illustrates this process with yet another example of a plate boundary—the Rift Valley in East Africa. What is now understood to be the emerging separation between two plates was instead accredited to a series of faults that created areas of uplift (block mountains) and subsidence (lakes).[62]

[60] Du Toit, *Physical Geography for South African Schools*, pp. 146, 148.
[61] Du Toit, *Physical Geography for South African Schools*, pp. 150, 158–159.
[62] Du Toit, *Physical Geography for South African Schools*, pp. 172, 176, 179.

With unerring instinct he then addressed the creation of mountains due to folding—a process that is now known to be a result of plate convergence and is evident in the Alps as the African plate moves towards the Eurasian plate and the Himalayas as the Indian plate collides rapidly with the Eurasian plate.

Reading *Physical Geography for South African Schools* is both provocative and frustrating. Written in far more accessible language than his academic work on continental drift, it is an illustration of Du Toit's incredibly perceptive thinking in relation to plate tectonics, his ability to detect the surface evidence that form the pieces of an elaborate jigsaw puzzle. It simultaneously is a demonstration of the inevitable criticism of continental drift theory on the part of Du Toit and earlier theorists—their necessary limitation in providing a convincing mechanism for crustal movement. This would have to wait for the technology that lay two decades in the future. Simultaneously, Du Toit's textbooks were more ambitious than a simple collation of existing secondary literature. He incorporated his own research and thinking, challenging his young readership and drawing them into the South African scientific nation.

Economic Geology and Mining

Du Toit's work in South America also had other unintended consequences. In his presidential address to the Geological Society of South Africa in 1928, Du Toit highlighted the real-world implications of drift theory. For a country as economically dependent on the extraction of mineral resources as South Africa, continental drift theory had economic implications and was more than an intellectual exercise. The correlation between South Africa and South America—in particular, Brazil—offered possibilities for economic geology through an understanding of the deposition processes of mineral resources such as manganese and diamonds. If minerals could be found on one continent why not the other?[63] Du Toit's remarks were seized upon by the press with an article in *The Star* emphasising the possibility that the kimberlite pipes of the Boa Vista mine in Minaas Geraes, Brazil were of an older age than the diamond-bearing pipes in South Africa that suggested the possibility that corresponding pipes in South Africa

[63] UCT-JL: Alex L. Du Toit Papers—BC 722: A4—Personal Documents: Alex Du Toit, "Some Reflections upon a Geological Comparison of South Africa with South America: Presidential Address delivered to the Geological Society of South Africa at the Annual Meeting, held on 19th March, 1928", p. 38.

were yet to be discovered. Du Toit had further emphasised the correlation between the coasts of Brazil and South West Africa that had both yielded the precious stones.[64] Du Toit's views had likely been influenced by the work of South African geologist, David Draper, who, on a trip to the Agua Suja Diamond Mine in Brazil in 1908, contradicted the Brazilian state geologist's view that Brazil's diamonds were of alluvial origin. Draper hypothesised that, like the diamonds of Kimberley, they originated in kimberlite pipes. A later trip also had Draper pay special attention to the parallels between rock types in South Africa and South America.[65] By highlighting the economic potential of continental drift theory, Du Toit was perhaps hoping to make his views more palatable to his South African colleagues yet the response by Professor R.B. Young who delivered the vote of thanks, pointed to the significant weakness of drift theory: "'So far as I know,' said Professor Young, 'no supporter of this theory has been able to point to any natural force sufficient to bring about this wonderful drifting apart of the continents'".[66]

Du Toit's comments were reported as far afield as the United States and a newspaper article sent to him suggested that a number of South Americans had seized upon the notion of continental drift with gusto. Termed "the greatest stampede Brazil has ever known", thousands of enterprising and prospective miners had begun searching for alluvial diamonds, creating almost overnight a gold-rush town named Lageado nicknamed the "Brazilian Kimberley". Their optimism was fuelled by the discovery of diamonds when water levels were low and the picture painted was of nineteenth century adventure that was far removed from the academic debates around continental drift: "the wilder aspects of frontier life have been revived; day and night, it is said, dance hall, cabarets, gambling halls and jazz bands are operating. Given the source of the 'rush', joined in by men who have no thought of contributing evidence to the drifting continent theory, and one of the most important 'missing links' between South America and South Africa may yet turn out to be the diamond".[67]

[64] UCT-JL: Alex L. Du Toit Papers—BC 722: M—Folders: M1—Book reviews/Biographies and Obituaries, 1926–1932. "American-African Link: The Evidence of the Diamond Fields" in *The Star*, 20 March 1928.

[65] Suryakanthie Chetty, "David Draper: The Making of a South African Geologist" in *Historia*, Vol 62, No 2, November 2018, pp. 35–36.

[66] "American-African Link: The Evidence of the Diamond Fields".

[67] UCT-JL: Alex L. Du Toit Papers—BC 722: M—Folders: M1—Book reviews/Biographies and Obituaries, 1926–1932. Newspaper clipping fragment sent to Du Toit by Reginald Daly—name and date unknown.

While research into continental drift made up a significant part of Du Toit's career, it was research he conducted while engaged in more "official" activities such as borehole location during the First World War or geological mapping in the first two decades of the twentieth century. This pattern continued when he moved into a new phase in his career. In 1927 Du Toit left government employment to join the private sector, becoming a geological consultant to De Beers Consolidated Mines.[68] The move to De Beers carried with it the hope of being free of the obligations in the employ of the state with the potential for carrying out the research which was clearly Du Toit's passion. In response to a tongue-in-cheek letter from J.D. Falconer of the Instituto de Geologia, Montevideo who wrote that Du Toit has succumbed to the "lure of the diamond", Du Toit's response was revealing of his prioritisation of research over material concerns (and the potential for personal accumulation associated with economic geology in South Africa):

> My abandonment of the Govt Service was due to the desire of becoming able to do research work before I get to [*sic*] old for travel, & to obtain funds towards that end, & I am very pleased my duties are not interfering with other scientific observations & that it is still possible to pay attention to other lines of geological investigation.[69]

Du Toit's association with De Beers related to yet another field—economic geology which relates to the extraction of minerals that can be put to use in industry—such as metals—or serve as an economic resource, including precious metals such as gold and platinum and minerals such as diamond. Economic geology in South Africa had been exemplified by the work of Hans Merensky. The formative years of Du Toit's early career in South Africa had also equipped him with the skills required for economic geology. His initial qualifications at the Royal Technical College in Glasgow were in Surveying and Mining Engineering. To this was added the fieldwork that detailed platinum and copper-nickel deposits in

[68] T.W. Gevers. "The Life and Work of Dr. Alex. L. Du Toit"—Alex L. Du Toit Memorial Lectures No. 1. In *The Geological Society of South Africa: Annexure to Volume LII* (Johannesburg, The Geological Society of South Africa, The South African Association for the Advancement of Science, The South African Geological Society, 1949), p. 27.

[69] UCT-JL: Alex L. Du Toit Papers—BC 722: B2—Correspondence with colleagues in South America (ca. 1920s–1931): Letter by JD Falconer to Du Toit, 6 February 1930 and Du Toit's response, 22 March 1930.

Griqualand East, the coal deposits that ran through Natal, the Transvaal, Swaziland and Zululand and the oil shale of Natal.[70]

Du Toit worked for De Beers until his retirement in 1941, first working in Kimberley before transferring to Johannesburg. He had already some familiarity with kimberlite pipes, originating within the earth's interior and a source of diamonds, as well as the alluvial source of diamond deposits that could be found in the Orange and Vaal Rivers. He was able to extend his research further afield, moving beyond South Africa to encompass the rest of the continent. It gives something of an inkling into Du Toit's character when he lamented the loss of knowledge of the geology of an area that had been subject to the ravages of mineral exploitation. For Du Toit, each piece of geological data was to be preserved in the hopes that someday it may be retrieved from its isolation and placed within a larger framework.[71]

Du Toit's work at De Beers gave him the opportunity for international travel including the various diamond-bearing regions on the African continent. Never satisfied with merely the practical and technological aspect of geology, Du Toit's academic publications considered mineral origins and distribution. Nor was continental drift neglected. Du Toit wrote extensively on fossilised flora from the Gondwana era found in Uganda, the glaciation periods in Antarctica and the formation of the sea floor. This culminated in *Our Wandering Continents* published in 1937. The opportunity for travel, from the United States to Russia and India, also allowed him to make and renew relationships with geologists all over the world and receive recognition from the international geological community.[72]

[70] Gevers, "The Life and Work of Dr. Alex. L. Du Toit", pp. 64–65.

[71] Gevers, "The Life and Work of Dr. Alex. L. Du Toit", pp. 27, 68–69.

[72] Gevers, "The Life and Work of Dr. Alex. L. Du Toit", p. 27.

CHAPTER 8

The Cradle of Humankind: A Pivotal Decade for Science in South Africa

The mid- to late nineteenth century saw a flurry of intellectual and scientific societies initiated in South Africa. This was temporarily halted by the South African War and then experienced a resurgence. Older societies were given new vigour and new organisations were established, one of which was the Southern Africa Association for the Advancement of Science (S_2A_3). First envisioned as an association for engineers, it was broadened to include scientists and modelled in part on the British Association for the Advancement of Science which had been formed 70 years earlier. This was explicitly set out at an early meeting held in Cape Town in 1901 to lay out the guiding principles of the S_2A_3: "That this meeting approves, and hereby confirms, the formation of a South African Association for the Advancement of Science, as far as possible on the lines of the British Association".[1]

David Gill who was in charge of the Royal Observatory at the Cape was named the first president of the S_2A_3 and members tended to belong to both the South African Philosophical Society (SAPS) as well as the S_2A_3. While the interests of both organisations often converged, the SAPS was usually more narrowly focused on the Western Cape with a membership

[1] Cornelis Plug, "The origin and early history of S_2A_3" in *Rudolf Marloth Brochure: Centenary Edition 1902–2002* Cornelis Plug, ed. (Boordfontein: Forum Press, 2002) p. 4. http://s2a3.org.za/joomla/files/archives/Marlth_Brochure_Centenary_Edition_2002.pdf, Accessed 2 December 2016, p. 5.

S. Chetty, *'Africa Forms the Key'*, Cambridge Imperial and Post-Colonial Studies, https://doi.org/10.1007/978-3-030-52711-2_8

composed almost exclusively of professional scientists. S_2A_3, on the other hand, had a regional membership, holding meetings as far afield as Maputo and Windhoek with a membership that was more inclusive and, at its height, reached 1300 in the wake of the first visit of the British Association.[2]

A highlight in the very early history of the S_2A_3 occurred in 1905 when the British Association for the Advancement of Science toured southern Africa, with a large contingent of 380 scientists. A book detailing the history of scientific progress in Africa, *Science in South Africa*, was published in anticipation of the visit and the papers delivered during the visit in sessions in Cape Town and Johannesburg were compiled in a four-volume collection. The president of the British Association was George H. Darwin, son of the eminent naturalist, who believed that the British visit would "[stimulate] scientific activity" in South Africa.[3]

Developments in the 1920s

The British Association for the Advancement of Science visited South Africa for the second time in 1929. Its members would find a country much altered politically, socially and economically to the one of 1905—and a country that had been imbued with a new sense of confidence. This changing relationship was evident in the introductory meeting of the two when, in the Cape Town City Hall, the president of the British Association, Sir Thomas Holland, "quoted in Afrikaans from Dr Viljoen's translation of Fitzpatrick's *Jock of the Bushveld*, adding: 'The translation is in a language capable of expressing virile sentences'".[4] This, too, would set the tone of the rest of the visit—an acknowledgement of the unique nature of South Africa but, one, which had its origins in British enterprise.

The difference between the two visits would have been particularly noticeable in intellectual and scientific endeavour. In many ways, the tour of the British Association marked the culmination of a decade of scientific discovery in South Africa that both challenged existing perceptions while being simultaneously shaped by a particular ideological milieu. This was evident in the fields of palaeoanthropology and geology.

[2] Plug, "The origin and early history of S_2A_3", pp. 5–6.

[3] Plug, "The origin and early history of S_2A_3", p. 6.

[4] "Scientists in S. Africa" in *The Times*, July 23, 1929, Issue 45262, p. 14. London, Times Newspapers Limited. Gale Document Number CS67301649.

1924 was a watershed. Professor Raymond Dart was employed by the University of Witwatersrand medical school when, in the summer of that year, a fossilised baboon skull was brought to his attention by a student, Josephine Salmons. The skull had been discovered at a mine owned by the Northern Lime Company in Taung, Bechuanaland (Botswana). The baboon skull was unusual in that there had been very little fossilised evidence of primates in sub-Saharan Africa. With the assistance of geologist friend, R.B. Young, Dart was able to obtain two crates of similar specimens unearthed during the limestone extraction which were subsequently delivered to him in Johannesburg. It was in one of these crates that Dart made the discovery that would transform the understanding of human evolution—a skull that was neither ape nor human but had the characteristics of both. Dart evocatively captures the moment of his discovery and the realisation of its implications:

> I stood in the shade holding the brain as greedily as any miser hugs his gold, my mind racing ahead. Here, I was certain, was one of the most significant finds ever made in the history of anthropology. Darwin's largely discredited theory that man's early progenitors probably lived in Africa came back to me. Was I to be the instrument by which his 'missing link' was found?[5]

Dart named the skull of the "Taungs baby" *Australopithecus africanus*[6] or "southern ape of Africa" and soon received early congratulatory messages from Jan Smuts who, while president of the South African Association for the Advancement of Science, had just lost his position as Prime Minister of the country. Smuts's words reflect his perception of the discovery as a means of highlighting the scientific importance of South Africa:

> Your great keenness and zealous interest in anthropology have led to what may well prove an epoch-making discovery, not only of far-reaching importance from an anthropological point of view, but also calculated to concentrate attention on South Africa as the great field for scientific discovery which it undoubtedly is.[7]

[5] Raymond Dart and Dennis Craig, *Adventures with the Missing Link*. (London, Hamish Hamilton Ltd, 1959) pp. 1–3.

[6] Dart and Craig, *Adventures with the Missing Link*, p. 22.

[7] Dart and Craig, *Adventures with the Missing Link*, pp. 36–37.

Dart's account demonstrates the way in which scientific discovery was contexualised within a nationalist vision that emphasised South Africa's centrality in scientific and intellectual circles. This was epitomised by Jan Smuts. At the same time, it also highlights the networks of empire. Raymond Dart was an Australian whose initial feelings of going to South Africa was of a man in "exile" in a scientific backwater and one entirely too old to live "a pioneer's life". After studying medicine and developing an interest in zoology at the University of Queensland, Dart's career was given a boost by the British Association for the Advancement of Science which went to Australia in 1914. It was here that he first came into contact with leading anatomist, Grafton Elliot Smith, an intellectual hero. It was thus hardly a difficult decision for Dart to make when, at the conclusion of the First World War, he was offered a position as a demonstrator for Elliot Smith at University College, London. While under Elliot Smith, Dart developed a new interest in anthropology as a result of the former's work on the Piltdown Skull—since revealed to be a hoax. He was then encouraged by Elliot Smith to apply for the post of chair of anatomy at the University of the Witwatersrand, despite some reticence on Dart's part. Anatomist Arthur Keith, who advised Dart on his application, later expressed reservations about Dart's suitability for the position—South Africa was largely religiously conservative whereas Dart considered himself a "Freethinker" and he further tended to challenge established views.[8] Dart's arrival in South Africa in 1923 confirmed his own worst feelings—the medical school was dilapidated and poorly organised and Dart himself was confronted by conservative colleagues who were less than enamoured with his Australian roots. For Dart, however, it was this "abysmal lack of equipment and literature" which led to him focusing his attention on anthropology, paving the way for the discovery he would make the following year.[9]

Raymond Dart's account also points to the interconnected history of palaeoanthropology and geology. As a child growing up in rural Australia, a favourite activity was searching for precious minerals during the course of which he was more likely to find archaeological artefacts. His great discovery was the result of limestone extraction in Bechuanaland and based

[8] Christa Kuljian, *Darwin's Hunch: Science, Race and the Search for Human Origins.* (Johannesburg, Jacana Media, 2016) p. 38.

[9] Dart and Craig, *Adventures with the Missing Link*, pp. 26–31, 33–34.

on the intervention of R.B. Young, a Scottish geologist. And, as a result of the Taung discovery, fossil samples found in limestone quarries in South Africa were sent to Dart's office for examination, highlighting the enthusiasm for and significance of the initial find.[10] After careful excavation, including the use of his wife's knitting needle, the skull of the Taung child was revealed on 23 December 1924. Less than two months later, Dart's findings were published in an article in *Nature* where he classified the Taung child as the hitherto unknown *Australopithecus africanus*—which was bipedal, used its hands for manipulation and, in spite of having a smaller brain than expected, was considered by Dart to be the link "between living anthropoids and man".[11]

The Taung discovery was not without its controversy. Chief amongst its detractors was a religious contingent who vehemently opposed Darwin's theory of evolution. Both Elliot Smith and Arthur Keith contested its significance in their response to Dart's article.[12] Anthropologists also tended to focus on hominin fossil finds in China and Indonesia, arguing for the importance of Asia rather than Africa as key to hominin evolution. This, too, according to Dart, was based on a particular ideological mindset, "Asia was the cradle of western, northern and eastern mankind. What dissonant squawkings were these from the puny South African infant at Taungs?"[13] For Dart, it was the challenges to obtaining food in the southern African environment that increased competition and drove the process of evolution.[14]

Simultaneously, however, even as Dart's findings challenged the centrality of Asia as the evolutionary homeland for humankind, he was less inclined to accord a similar importance to contemporary indigenous Africans. Addressing the South African Association for the Advancement of Science in 1925, Dart emphasised the importance of the uniquely South African—and African—context of scientific study yet considered science to be a hallmark of civilisation, one that should not solely be confined to Europe but incorporate more marginalised spaces such as South Africa. Yet, Dart's view of the African researcher was a narrow one—scientific

[10] Dart and Craig, *Adventures with the Missing Link*, pp. 26, 3, 55.

[11] Kuljian, *Darwin's Hunch*, p. 43.

[12] Saul Dubow, "Human Origins, Racial Typology and the Other Raymond Dart" in *African Studies*, 55:1, 1996, p. 2.

[13] Dart and Craig, *Adventures with the Missing Link*, pp. 43, 52–54.

[14] Kuljian, *Darwin's Hunch*, p. 44.

study was the prerogative of whites with blacks forming the subject of research.[15] Dart had also made a study of the ruins of Great Zimbabwe, believing as had others before him, that the ruined settlement had not been a result of African agency but was, instead, a remnant of a Mediterranean civilisation that had dominated southern Africa. The influences of this invasion could even be seen in the ancient cave paintings that dotted the region. Evidence at the site of clearly African origin were discounted by early archaeologists and dismissed by Dart as simply "Bantu contamination". The exhaustive excavations of British archaeologist Gertrude Caton-Thompson challenged this view held by Dart. In a paper presented in Johannesburg to the British Association for the Advancement of Science during its visit to South Africa in 1929, Caton-Thompson demonstrated that the lay-out of Great Zimbabwe originated in the indigenous kraal structure which, along with the wealth of evidence so easily dismissed previously, demonstrated conclusively that Great Zimbabwe was of indigenous African origin. She reinforced this by leading the Association on a tour of the ruins. Dart, however, was not so easily persuaded and was, in fact, outraged by her conclusion. In this he was supported by many white South Africans and Rhodesians alike.[16] This vignette illustrates the ways in which scientific evidence could be harnessed to ideology. Du Toit's argument for African centrality compounded by Dart's palaeoanthropological discoveries rendered the African continent significant in a manner that did not threaten colonial domination of the continent and also served to strengthen settler nationalism. The findings at Great Zimbabwe, however, could not be credited to a distant prehistoric people and had the potential to challenge arguments for racial superiority and political dominance held by white settler populations.

Sub-Saharan Africa's centrality in hominin evolution would be greatly strengthened by further discoveries. One of Dart's congratulators—and supporters—was Dr Robert Broom who would go on to achieve his own fame with his finds in Sterkfontein.[17] The Sterkfontein region, known as the "Cradle of Humankind", in Gauteng, South Africa is a product of its geology. Composed largely of dolomite—magnesium carbonate—which

[15] Dubow, "Human Origins", pp. 7–8.

[16] Michael F. Robinson, *The Lost White Tribe: Explorers, Scientists, and the Theory that Changed a Continent*. (New York, Oxford University Press, 2016) pp. 204, 208–210.

[17] Kuljian, *Darwin's Hunch*, pp. 47, 84–85.

reacted with water to form extensive cave formations, the region became the abode of prehistoric predatory cats and thus a site for bone deposits of both predator and prey. The caves were later also a natural source of shelter for early hominins and subsequent site for their remains. Due to its geology, then, Sterkfontein, Swartkrans, Kromdraai and other sites in the region maintain a unique record of hominin evolution.[18] It seems fitting then that a geologist was linked to the first fossil finds of the region in 1895. This was the year that Hans Paul Thomasset began exploiting the limestone deposits of the area. On the 1st of February a box of assorted pieces of fossilised bone was sent to the British Museum of Natural History by David Draper, one of the foremost geologists in South Africa. Although Draper had yet to visit the cave from which the samples were taken, three months later he presented a paper before the South African Geological Society detailing landscape formation as a result of dolomite and spoke of the formation of cave systems that served as a repository for animal remains. In 1897, he made public the discovery of a large cave system in the region.[19]

David Draper was born in the Cape in 1849, the son of an 1820 settler who was employed as a botanist by the Cape Government Service. His fascination with geology began with the discovery of diamonds in Kimberley and, while still a teenager, Draper travelled to the Kimberley diamond fields and, later to the gold deposits in Barberton and the coal deposits in Natal, eventually becoming a geologist. Interestingly though, his geological expertise was derived from this practical experience as well as his encounters with some of the prominent geologists of the time such as Andrew Geddes Bain. Draper himself did not complete a formal academic education as a geologist but this had little adverse effect on his career and he was the first geologist born in South Africa to become a Fellow of the Geological Society of London. One of the papers he presented to this institution was "The Occurrence of Dolomite in South Africa" in 1894, predisposing him to understand the formation of the Sterkfontein cave system. The growing exploitation of mineral deposits in

[18] Philip Bonner, "Africa is Seldom What It Seems" in *A Search for Origins: Science, History and South Africa's 'Cradle of Humankind'*, Philip Bonner, Amanda Esterhuysen and Trefor Jenkins, (eds). (Johannesburg, Wits University Press, 2007) pp. 2–3.

[19] Phillip V Tobias, "The Story of Sterkfontein Since 1895" in *A Search for Origins: Science, History and South Africa's 'Cradle of Humankind'*, Philip Bonner, Amanda Esterhuysen and Trefor Jenkins, (eds). (Johannesburg: Wits University Press, 2007) pp. 216–220.

South Africa suggested to Draper the need for a local body for geologists in the same vein as the Geological Society. It was therefore largely through the efforts of David Draper and Hugh Exton that the Geological Society of South Africa (GSSA) was formed in 1895. Its first meeting took place in the Council Room of the Transvaal Chamber of Mines, highlighting the close links between geology and the mining industry. Draper was appointed the first Secretary and Treasurer of the GSSA, with Hugh Exton the first president.[20] In his address before the society in February 1898, Exton would also discuss the geology of the Sterkfontein caves in detail.[21]

The true value of Sterkfontein for understanding hominin evolution would arguably only begin decades later. Dart's discovery at Taung paved the way for a new understanding of Africa's place in hominin evolution and Robert Broom would bear this out in 1936. In August Broom who was employed by the Transvaal Museum made the first mature *Australopithecus* fossil find—and this would become the first of many. Sterkfontein, Kromdraai and Swartkrans provided—and continue to provide—fertile ground for the story of human evolution. Du Toit's perception of Africa being "key" to an understanding of the geological past with the publication of *Our Wandering Continents* in 1937 would have its antecedents in Broom's claim a year earlier: "I feel sure that within a very few years the whole problem of man's origin will be solved and be solved in Africa".[22] By the Second World War, East Africa also provided growing evidence for Africa's centrality in human evolution with fossil evidence obtained from Kenya and Ethiopia.[23]

Alex Du Toit was not isolated from these developments in palaeoanthropology and archaeology. His extensive early field work had exposed him to the stone tools as well as rock paintings found in the Stormberg and Drakensberg regions and Du Toit believed that the fields of

[20] "Dr David Draper (1849–1929)" in *A Century of Geological Endeavour in Southern Africa, 1895–1995*, C.R. Anhaeusser, (ed). (Linden, The Geological Society of South Africa, 1997) p. 24.

[21] Tobias, "The Story of Sterkfontein since 1895", p. 222.

[22] UCT-JL: Alex L. Du Toit Papers—BC 722: F3—News clippings and Articles: F3.1—Vaal River Prehistoric Fossil Finds: "Discovery new Krugersdorp: Dr Broom's Finds" in *The Star*, 25 August 1936.

[23] Phillip V. Tobias, "A Century of Research in Human Biology and Palaeo-anthropology in Southern Africa" in *A History of Scientific Endeavour in South Africa: A collection of essays published on the occasion of the Centenary of the Royal Society of South Africa*, A.C. Brown, (ed). (Cape Town: The Royal Society of South Africa, 1977) pp. 231–233.

archaeology and geology were complementary with the latter needing to include an assessment of human effects on environment and both disciplines essential for stratigraphic correlation. Du Toit was also a member of the Council of the South African Archaeological Society, becoming its second president in 1946. His seminal *Geology of South Africa* also addressed hominin evolution in Southern Africa in a chapter entitled "Primitive Man".[24] In Du Toit's work, at least, therefore there was a convergence between his geological thinking and other developments related to prehistoric southern Africa. Du Toit's documents contained numerous articles detailing the work of Broom and Dart, demonstrating his interest in prehistoric South Africa. He was, however, called upon to play a more active role, bringing together the fields of geology and palaeoanthropology.

Gravel is composed of fairly coarse pieces of rock that are eroded and can be transported by water action. Easily transported when rivers are fast flowing and currents are stronger, they are deposited as the current decreases in strength. This leads to the formation of gravel beds which are indicative of the past course of a river. During the course of his geological work, Du Toit had described the gravel beds of the Vaal River and had subsequently prepared a report on the gravels, "Note on the earlier Gravels of the Vaal River between Barkly West and Windsorton", with accompanying descriptions and possible explanations for the means by which they were deposited as well as the implications for understanding past climate. His work had demonstrated that there had been three gravel beds, all evidence for the existence of water flow and this took on new significance with Dart's work.[25]

Near Bloemhof, a town located along the Vaal River, Dart found evidence of prehistoric life in the form of the teeth of mammoths, the first found in the southern hemisphere. The gravels also contained further evidence of prehistoric human activity with stone tools which, like the teeth, did not display the characteristic smoothening brought about by river transportation, suggesting that they were found where they were first

[24] T.W. Gevers, "The Life and Work of Dr. Alex. L. Du Toit"—Alex L. Du Toit Memorial Lectures No. 1. In *The Geological Society of South Africa: Annexure to Volume LII*. (Johannesburg, The Geological Society of South Africa, The South African Association for the Advancement of Science, The South African Geological Society, 1949) pp. 76–78.

[25] UCT-JL: Alex L. Du Toit Papers—BC 722: F3—Newspaper Clippings and Articles: F3.1 Vaal River Prehistoric Fossil Finds.

deposited. The site was thus considered to be an early site of prehistoric human habitation with Dart's fervent wish that it would yield more hominin discoveries along the lines of the Taung skull.[26] The effects of the find were to lead to a rethinking of the age of human habitation in South Africa—that would correspond to the findings of Dart and Broom. Moreover, as stated by one of the excavators, Van Riet Lowe, the discoveries had allowed South Africa to claim an antiquity as old as that evident in the northern hemisphere: "This sequence of cultures, this gradual evolution of human development, corresponds exactly with that already proved for Europe and North Africa, and it is illuminating to know that we have remains of man's handiwork here that are comparable with and culturally as old as any that have been recognised in other parts of the world". At the same time, Van Riet Lowe took care to differentiate these early inhabitants from the later "Bush folk" as the former were conveniently extinct.[27] Like Dart, then, it was important to clearly distinguish ancient humans in southern Africa from the contemporary inhabitants with land occupation being a charged issue in a climate of segregation and dispossession.

The initial discovery only related to the uppermost (and hence most recent) gravel bed and more finds were made prompting Dart to assert the unique trend of South Africa's evolutionary history that owed little to Europe and North America yet was of comparable age: "[The mammoth teeth] showed that South Africa had her own local history of elephant evolution and her own types which had not migrated from Europe, India or America. They revealed the enormous period of time that had been necessary to the production of these types".[28] In the wake of these discoveries, Dart was therefore pleased to accept Du Toit's offer to continue geological exploration of the region with an emphasis on determining the "antiquity" of human occupation as derived from a study of the gravel beds, with particular focus on the lower regions. Du Toit's efficacy would

[26] UCT-JL: Alex L. Du Toit Papers—BC 722: F3—Newspaper Clippings and Articles: F3.1 Vaal River Prehistoric Fossil Finds: "Africa as Grave of Primitive Man: Full Story of Important Bloemhof Discovery" in *Rand Daily Mail*, 1 November 1927.

[27] UCT-JL: Alex L. Du Toit Papers—BC 722: F3—Newspaper Clippings and Articles: F3.1 Vaal River Prehistoric Fossil Finds: "The Mammoth-Hunters of the Vaal: Striking Scientific Discovery: First Proofs of South African Stone Age Theory" in *The Star*, 8 December 1928.

[28] UCT-JL: Alex L. Du Toit Papers—BC 722: F3—Newspaper Clippings and Articles: F3.1 Vaal River Prehistoric Fossil Finds: "Africa the Home of Elephants: Prof Dart Traces Evolution" in *The Rand Daily Mail*, [date unknown], 1929.

be in giving a broader picture of the area as a whole for the archaeologists and palaeoanthropologists to envisage early human settlement and utilisation of the landscape.[29] Geology was thus to be harnessed to the project of determining human origins—and South African significance. These would be themes that Du Toit would himself mention, arguing for an antiquity that predated that of humankind itself.

In 1924 then, contextualised by new developments and discoveries in South Africa that challenged existing understandings of the history of humankind, by Smuts's Holism and by Du Toit's own hypotheses regarding the history of the Earth, Alex Du Toit presented a paper that highlighted the unique nature of South Africa's geology. This remarkable presentation was clearly an assertion of South Africa's unique nature that set up the autonomy of the fledgling nation against that of established Europe and confirmed its pre-eminence in the geological history of the Earth. Du Toit's presentation therefore begins with a criticism of the Eurocentric and parochial nature of the discipline of geology.

For Du Toit, the origins of modern geology in Europe and the preponderance of research undertaken in the northern hemisphere created a dogmatic view of the Earth's processes, making the scientists of the northern hemisphere unwilling or unable to accept new hypotheses. This hostility to alternative views did not allow for new understandings and new discoveries that inevitably arose as more of the world was subject to geological investigation. With its European focus, the discipline of geology had developed "a somewhat narrow outlook on what are proving to be really world-wide problems".[30]

This understanding was obsolete in the light of the expansion of modern scientific investigation and research beyond the borders of its inception. With areas outside Europe now being opened up to geological study, it was clear to Du Toit that new evidence necessarily had to challenge and adapt existing conceptions of the Earth's history. Geology was no longer a solely European enterprise with each part of the world being able to make a meaningful contribution and, by so doing, have the potential "to influence and often radically to modify geological theory and hypothesis".[31]

[29] UCT-JL: Alex L. Du Toit Papers—BC 722: F3—Newspaper Clippings and Articles: F3.1 Vaal River Prehistoric Fossil Finds: Letter from Raymond Dart, University of the Witwatersrand to Miss A. Wilman, 22 October 1928.

[30] Alex L. Du Toit. "The Contribution of South Africa to the Principles of Geology" in *South African Journal of Science*, Vol. 21, pp. 52–78, November 1924. http://www.sabinet.co.za Doc AJA00382353_2525, Accessed 13 October 2016, p. 52.

[31] Du Toit, "The Contribution of South Africa to the Principles of Geology", p. 52.

With his opening salvo Du Toit was asserting the agency and capability of the formerly colonised against the metropole, scientific—and specifically, geological—knowledge forming the terrain of the confrontation.

Du Toit's paper is a challenge to the unidirectional transmission of knowledge from the northern hemisphere—the centres of Enlightened learning—to the south. It suggests a frustration at the dismissal of knowledge created in these "outposts", at their failure to recognise and accord equality:

> There is hardly a branch of geology that has not beneficially been touched and advanced by investigation in South Africa, although a great deal of such valuable work, as is only natural, has come to be overlooked abroad, or the bearing thereof insufficiently appreciated.[32]

Du Toit's apparent exasperation occurs, not only within an assertion of the scientific capability of South Africa, but as a means of asserting South African dominance on the African continent itself. He draws attention to the rest of southern Africa, beyond the borders of the Union, which require greater penetration by a new generation of explorers—not motivated by gold, glory and adventure—but in the interests of geology, geography and botany, in the desire to know and to subject these areas to the gaze of scientific investigation.[33]

He goes on to highlight the nature of South Africa's geology—its stratigraphy, petrology and with, unsurprisingly, a particular focus on its mineralogy. Two aspects of his discussion are of particular interest in relation to continental drift. The first is kimberlite pipes, the source of diamonds, which Du Toit believed were an indication of volcanic activity arising from "rifts or fracture lines situated at quite moderate depths within the crust". He also mentions zircons found in ancient granite in Mozambique which studies suggested was 1.5 billion years old and Du Toit hoped that similar studies could be carried out on South African rocks.[34] His brief reference to zircons is remarkably prescient as it is the analysis of this mineral that currently provides insight into tectonic activity.

[32] Du Toit, "The Contribution of South Africa to the Principles of Geology", p. 53.
[33] Du Toit, "The Contribution of South Africa to the Principles of Geology", p. 53.
[34] Du Toit, "The Contribution of South Africa to the Principles of Geology", pp. 58–59.

The surface of the Earth is constantly changing with both oceanic and continental crust recycled and the latter subject to weathering and erosion. Zircon ($ZrSiO^4$) is a mineral that is extremely durable and is able to survive these destructive geological processes.[35] Found largely in the granitic rock of which continental crust is composed, zircons may be billions of years old and are able to indicate ancient geological activity, including that of tectonics.[36]

In terms of both its economic significance as a site of mineral deposits as well as being an example of South Africa's unique geology, Du Toit draws attention the Bushveld Igneous Complex which is the most extensive geological feature of its kind anywhere in the world.[37] The Complex is composed of layers of igneous rock that, at their maximum, reach almost eight kilometres in width. The Complex itself covers an area of approximately 65,000 square kilometres.[38] For Du Toit, the sheer size of the Complex and its distinctive stratigraphy would make it a "classic" area of study in terms of offering insight into igneous intrusions, the process of mineralisation and the formation of metamorphic rock due to the high temperatures of the igneous material altering the structure of the surrounding country rock.[39]

Whereas the Bushveld Igneous Complex offered the potential to understanding the interaction between heat and rock, the Dwyka Conglomerate provided an unparalleled understanding of glaciation—and raised some interesting questions regarding continental movement. With Du Toit's particular expertise in glaciation, his discussion of the Dwyka Conglomerate forms the prelude to a discussion on Gondwana. He indicates the similarities of this Conglomerate with the Talchir Conglomerate in India and goes on to highlight similar evidence of glaciation in Madagascar, South

[35] Nicola Jones. "Minerals yield signs of early plate tectonics" in *Nature*, 26 November 2008. http://www.nature.com/news/2008/081126/full/news.2008.1256.html, Accessed 4 May 2017.

[36] "Researchers confirm the existence of a 'lost continent' under Mauritius", 31 January 2017. https://phys.org/news/2017-01-lost-continent-mauritius.html, Accessed 4 May 2017.

[37] Du Toit, "The Contribution of South Africa to the Principles of Geology", p. 64.

[38] R.G. Cawthorn, H.V. Eales, F. Walrave, R. Uken and M.K. Watkeys, "The Bushveld Complex" in *The Geology of South Africa*, M.R. Johnson, C.R. Anhaeusser and R.J. Thomas, (eds). (Johannesburg and Pretoria, The Geological Society of South Africa and the Council for Geoscience, 2006) p. 261.

[39] Du Toit, "The Contribution of South Africa to the Principles of Geology", p. 64.

America, the Falkland Islands and Australia. Yet it is glaciation in South Africa that provides the vital evidence: "there is no doubt that South Africa has provided information concerning this climatic revolution of exceptional importance". What makes South Africa particularly significant here is that it shows the ice moving towards the South Pole rather than originating from a southerly direction. Furthermore, four different periods of glaciation can be discerned in the geological record indicating different and varying "centres of glaciation" where the ice moved in unexpected directions. These uncharacteristic movements in relation to the South Pole can only be accounted for by either suggesting that the poles moved, "polar wandering", or that the continents did. For Du Toit, South Africa (and he uses "South Africa" interchangeably with "Africa") was strategically placed to experience these recurrent glacial events and thus is of extreme importance in studying continental drift.[40]

South Africa's geographical location is just as relevant when Du Toit subsequently addresses "Gondwanaland". As well as blithely dismissing all criticism and considering the former unity of the southern continents a foregone conclusion, he argues for South Africa's centrality within the supercontinent itself, heightening its prominence in relation to the other landmasses: "South Africa seems to have formed almost at the centre, and by virtue thereof ranks rather higher in importance than the other components". This is affirmed by the variety of fossil and stratigraphic evidence found in South Africa—which has been addressed elsewhere in this book. And South Africa's centrality once again comes to the fore when Du Toit considers the subsequent break-up of Gondwana. He creates a hierarchy of southern landmasses based on the evidence they have to offer—at the apex is South Africa, followed closely by South America with "Australia [forming] a rather remote corner of Gondwanaland" and India "adding relatively little to our knowledge of late". When one looks at a contemporary map of the world, the origins of the map are often clear. If created in the United States, the North and South American continents are centrally situated; if made in Europe, Eurasia and, below it, Africa form the centre of the map. In Du Toit's recreation of Gondwana, South Africa became emblematic of Africa as a whole and was allocated a starring role in the reconstruction and subsequent break-up of Gondwana. The very tip of the African continent, described as "this truly wonderful land of ours" provided the essential evidence by which an ancient supercontinent could

[40] Du Toit, "The Contribution of South Africa to the Principles of Geology", pp. 67–69.

be determined and, in so doing, challenge the geological hegemony of the northern hemisphere:

> The more outstanding instances [of South Africa's unique geology] are already finding their way into the newer text-books and it will not be long before some of them will become familiar to scientists abroad, which in turn would encourage the small band of workers in this Continent to further and more energetic research notwithstanding the disadvantages under which they are labouring.[41]

In his conclusion, Du Toit thus paints a compelling picture of the complexities by which geological knowledge is acquired, shaped, adapted and challenged, defying the unidirectional transmission of knowledge from north to south, from the metropole to the former colony and new nation.

Just as hominin fossil remains suggested greater importance for the African continent in human prehistory, challenging existing scientific thought that prioritised the East, "continental drift" theory would literally upset the ground on which these scientists stood.

Wegener's hypothesis on "continental drift" was first translated into English in 1924, drawing the largely hostile attention of the English-speaking world. Forty years before plate tectonics would gain scientific credibility, the movement of continents was accepted by Du Toit with the wholehearted support of Smuts. Just a year later, in his capacity as president of the South African Association for the Advancement of Science, Jan Smuts spoke of a particularly South African vision of science that would challenge the "habits of thought and the viewpoints characteristic of its [science] birthplace in the northern hemisphere". In his address, Smuts highlighted the way in which recent hominin fossil discoveries gave the country a central place in the story of human evolution. At the same time "continental drift" had placed Africa at the geographical centre of the southern hemisphere. Africa was the "mother continent", the central part of a large landmass from which the other continents—Australia, India, Madagascar and South America—had fragmented.[42] Both fields also bolstered Smuts's own philosophy of Holism which was reaching its final stages of development.

[41] Du Toit, "The Contribution of South Africa to the Principles of Geology", pp. 69, 72, 77.

[42] Saul Dubow, "White South Africa and the South Africanisation of Science: Humankind or Kinds of Humans?" in *A Search for Origins: Science, History and South Africa's 'Cradle of Humankind'*, Philip Bonner, Amanda Esterhuysen and Trefor Jenkins, (eds). (Johannesburg, Wits University Press, 2007) pp. 10, 12.

1924 was a significant year for Jan Smuts. His South African Party was defeated in the election and a new Pact government came into place, composed of an alliance between the National and Labour Parties. Smuts stepped down as Prime Minister to be replaced by JBM Hertzog. During this enforced hiatus from a long and eventful political career, Smuts was able to finally pay attention to his scientific and philosophical interests whilst ensconced on his farm, Doornkloof, outside Pretoria. His particular field of expertise was botany and Wegener's hypothesis provided a useful means of understanding the origins of South Africa's tropical flora. The prevailing view was that the tropic flora had originated in northern latitudes; however, Smuts recalled a notion put forward by Darwin in the late nineteenth century in which Darwin posited the existence of an ancient southern continent which was the source of this tropical flora. Wegener's Gondwanaland seemed to lend credence to this view.[43] These early—and controversial—developments in geology (and palaeoanthropology) not only found a receptive audience in Smuts due to his scientific inclinations, they also formed part of the context of his philosophical development to which he would devote the following year.

Smuts had been flirting with a philosophy of "wholeness" or unity for decades and, in 1910, sought the advice of friend on a manuscript entitled *An Inquiry to the Whole* which he was advised to revise extensively.[44] Freed from the burdens of leadership, he returned to the concept and conceived the structure of *Holism and Evolution* in September 1924. The manuscript was completed in September 1925 and marked the culmination of Smuts's lifelong thinking on no less a subject than the universe and our place in it.[45] In *Holism and Evolution* Smuts contended with what he saw as a linear progression where the evolution of the universe took place according to a blueprint initiated at its creation. For Smuts, this was stifling:

> All real novelty and initiative, all real freedom of choice and development disappear from the universe. The process of the world becomes at most an explication, an unfolding of what was implicitly given, and not a creative evolution of new forms.[46]

[43] W.K. Hancock, *Smuts 2: The Fields of Force, 1919–1950*. (Cambridge, Cambridge University Press, 1968) pp. 164–165, 174–175.

[44] Richard Steyn, *Jan Smuts: Unafraid of Greatness*. (Johannesburg and Cape Town, Jonathan Ball Publishers, 2016) p. 203.

[45] Hancock, *Smuts 2: The Fields of Force*, pp. 176–178.

[46] J.C. Smuts, *Holism and Evolution*. (New York, The Macmillan Company, 1926) p. 89.

His view of the universe attempted to harmonise science with his own religious upbringing. Smuts drew upon physics and chemistry, biology and philosophy to argue for the creativity that stemmed from evolution and change. Beginning with the inanimate world at the level of the atom—the basic building block of all matter—to the world of plants, animals and, at the apex, human beings, Smuts argues that all are part of a progression towards ultimate perfection, the whole. While bearing a superficial resemblance to the Great Chain of Being, the process that drove evolution—particularly on the part of animate beings—was the voluntary action and freedom of beings to interact with and shape the path of their development. The ultimate liberation was human consciousness where human beings shaped their destiny, comprehended their world and brought into being the world of the Spirit, the realm of philosophy and religion.[47] In this way, Smuts hoped to unite the physical and the spiritual in the ultimate "whole".

In Gevers' biography of Alex Du Toit, he repeatedly describes the concept of continental drift as being "holistic" which is a reflection of Du Toit's own beliefs evident in *Our Wandering Continents* in 1937: "Indeed it is modestly suggested that the Displacement Hypothesis represents the Holistic outlook in Geology". Related to this—as with Smuts's conception—is the image of "evolution"—the progression and transformation of geological activities as part of a greater pattern. In Du Toit's vision, each piece could not be taken in isolation—they came together, interacting and shaping each other until a harmonious whole was created. The aesthetic nature of it was in the "order" it created. His description is poetic and would have resonated with Smuts: "Instead of such units behaving more or less arbitrarily they become under our hypothesis parts of a living whole, each influencing and reacting upon its neighbours in a definite and orderly fashion".[48]

The domination of scientific thought with its origins in the northern hemisphere was itself limiting. Modern scientific thought had developed in the north but it need not be confined there and the hominin finds suggested a new focus on the scientific contributions of the southern hemisphere. It allowed for the assertiveness and evolution of a distinct southern identity that was not merely a product of northern domination.

[47] Steyn, *Jan Smuts: Unafraid of Greatness*, p. 205.

[48] Gevers, "The Life and Work of Dr. Alex. L. Du Toit", p. 82.

Wegener's—and Du Toit's—continental drift merely strengthened this. Gondwanaland offered new possibilities for research; it was a means of understanding the interconnectedness of hitherto separate landmasses and the process of the movement and evolution of life. From the disparate elements would emerge a whole.

Du Toit's geological work was further drawn to Smuts's attention by Illtyd Buller Pole Evans. Pole Evans was born in Wales, the son of a clergyman who trained in botany at Cambridge University. He was subsequently employed by the Department of Agriculture in Pretoria in 1905, eventually heading the Division of Mycology and Plant Pathology in the Department of Agriculture after the formation of the Union of South Africa. Pole Evans shared Smuts's passion for botany—with a particular interest in fungi—and harnessed science to development and ecology through his work on plant diseases as well as the detrimental effects of grazing on vegetation. Eventually, he, Smuts and botanist, John Hutchinson, travelled through Africa reaching Lake Tanganyika. He became president of the South African Association for the Advancement of Science in 1922.[49] Pole Evans had also attended a debate entitled "The Nature of Life" between Smuts and Lancelot Hogben, Professor of Zoology at the University of Cape Town. The latter represented a mechanistic view of nature in contrast to the former's Holism. The mechanistic model focused on the individual parts rather than the whole and Pole Evans was firmly in the holism camp.[50] In a letter to Smuts in 1927, Pole Evans wrote glowingly of Du Toit's presentation on the Kalahari to the Biological Society and expressed regret at Du Toit's leaving government employ:

> I think it is a very great pity that he is leaving the Govt Service as you have probably heard. He evidently thinks the future does not hold out much prospects for the scientific man. When I get a few minutes to spare I have been reading his latest book in Geology. It is certainly the best thing of its kind that has yet been published on South Africa. He has an enormously wide grasp of his subject.[51]

[49] C. Plug, "Pole-Evans, Mr Illtyd Buller" in *S2A3 Biographical Database of Southern African Science*. http://www.s2a3.org.za/bio/Biograph_final.php?serial=2224, Accessed 7 February 2018.

[50] Peder Anker, "The Politics of Ecology in South Africa on the Radical Left" in *Journal of the History of Biology*, Vol. 37, No. 2, Summer 2004, pp. 310–311.

[51] UCT-JL: Smuts Papers Vol. 34, No. 122: Letter from IB Pole Evans to Smuts, 30 March 1927.

[52] C. Plug, "Pole-Evans, Mr Illtyd Buller".

Pole Evans was most likely referring to Du Toit's comprehensive *The Geology of South Africa* published previously and Pole Evans would himself work on a less ambitious survey of "The vegetation of South Africa" two years later for the visit of the British Association for the Advancement of Science.[52]

Smuts would eventually be given the position of president of the British Association for the Advancement of Science in 1931, indicating an acknowledgement of the development of intellectual endeavour in the former colony which had now literally placed itself in the centre of the world.[53]

The Second British Association Visit

Jan Hofmeyr was the president of the Southern African Association for the Advancement of Science when the British Association returned to the country in 1929. For most of Hofmeyr's political career, he was associated with the ideology exemplified by Smuts. Born in 1894, Jan Hendrik Hofmeyr was part of an illustrious Afrikaner family that included "Onze Jan" who would help establish the Afrikaner Bond.[54] The elder Hofmeyr had envisaged an organisation that—in a far cry from the Afrikaner nationalism that would come later—would forge a common South African identity between English and Afrikaans-speaking white South Africans who, while maintaining loyalty to the British Empire, would not simply be subservient to it. Afrikaners would still retain their cultural characteristics, epitomised by language, but would also benefit from the association with the metropole and British colonies. By the late nineteenth century, men such as Smuts were attracted to the ideals of the Afrikaner Bond.[55] This both contradictory and complementary relationship between nation and empire was therefore highlighted by the British Association visits.

The young Hofmeyr demonstrated great intellectual ability from very early on, learning to read in both Dutch and Afrikaans at the age of five. He was a child prodigy who began studying at the South African College at the age of 12 and graduated with distinction at the age of 15. For his efforts he was awarded a Rhodes Scholarship to study at Oxford University.

[53] Steyn, *Jan Smuts: Unafraid of Greatness*, p. 213.

[54] Alan Paton, *Hofmeyr*. (Cape Town, Oxford University Press, 1964) pp. 1–2.

[55] Herman Giliomee, *The Afrikaners: Biography of a People* (Cape Town, Tafelberg Publishers, 2003) pp. 221–222.

By 1919, Hofmeyr had begun to formulate his ideas regarding South Africa, both as an independent nation but also as part of an international community that had been shaped by British imperialism. He met Smuts a year later and was subsequently recommended by the statesman for the presidency of the South African Association for the Advancement of Science.[56] Hofmeyr's address in 1929 was thus contextualised by a particular nationalist and, in the case of Smuts, imperialist vision of South Africa. At the same time, he was approvingly described by the British press as having had "experience of the distinctive culture of an English university".[57]

Hofmeyr welcomed the British contingent with a speech entitled "Africa and Science". During the course of the oration, Hofmeyr employed the metaphor of parent and child in relation to the British and Southern African Associations respectively. The tone adopted was of a child addressing a venerable parent, "this stripling association brings its tribute of respectful admiration and good-will".[58]

While the earlier British visit in 1905—to which Hofmeyr alluded—had stimulated scientific interest, the intervening decades were a testament to the great strides made by South Africa in science and, associated with it, progress. Hofmeyr detailed developments in South Africa that included a growth in universities and tertiary education, a growing number of scientific organisations, a three-fold increase in astronomical observatories, developments in veterinary science and medicine and, of course, the recent palaeoanthropological discoveries that were challenging existing views of human origins.[59] These developments were also published in a book to commemorate the occasion of the visit in 1929 and provide a concrete record of South Africa's development for the five hundred delegates representing the British Association.[60]

But, to continue the metaphor, children have to attain a certain independence from their parents and develop their own identity. For Hofmeyr, this could be summed up as the "South Africanisation" of science. This

[56] Paton, *Hofmeyr*, pp. 3, 16, 89, 163.

[57] "Science in Africa" in *The Times*, July 23, 1929.

[58] Jan H. Hofmeyr, "Africa and Science" in *Science*, Vol LXX, No. 1812, Friday, September 20, 1929, p. 270.

[59] Hofmeyr, "Africa and Science", pp. 271–272.

[60] Plug, "The origin and early history of S_2A_3", p. 6.

referred to science from a South African perspective and focused on South African concerns—and was described rather vaguely as "the distinctive features of the South African outlook—freshness and breadth of view, receptivity to new illuminations and readiness to see old truths in new settings and in the light of their wider bearings". For Hofmeyr this could be more concretely exemplified in Jan Smuts's work on Holism, an example of the unique and global contribution made by South African scientific thought. Hofmeyr went on to quote Smuts's own address while president of the S_2A_3 four years previously when Smuts referred again to the "South African view-point" as bringing something of value to bear on scientific research. While, again, not explicitly spelled out, it was clear that South Africa was not merely following in the footsteps of the metropole but asserting a particular form of nationhood: "We are proud of our South African science not least because we know that we can regard it as distinctively ours".[61] In the British press, this was reported rather poetically: "The seeds brought from Europe have taken root in a new soil, have produced luxuriant foliage and blossoms and fruits different in flavour from those of the original plant".[62]

As South Africa did not yet exist as a political state when the S_2A_3 was formed, the organisation claimed to represent the southern African states which were largely white settler or colonial states in the early twentieth century.[63] By 1929, Hofmeyr envisaged South African science—with its roots in a European tradition—providing intellectual leadership for the African continent as a whole: "If, then, South Africa aspires to leadership in Africa in other branches of activity, why not also in science?"[64] This implied a form of hierarchy—Britain, South Africa and the rest of Africa. Its roots can be found in Smuts's political views regarding the position of South Africa on the African continent. Even prior to the unification of South Africa, Smuts envisioned the control of South Africa over neighbouring African regions with particular focus on Rhodesia as well as Mozambique. This, too, was a manifestation of holism—South African dominance in Africa within the ambit of a British Commonwealth.[65] Also

[61] Hofmeyr, "Africa and Science", pp. 272–273.

[62] "Science in Africa" in *The Times*, July 23, 1929, Issue 45262, p. 15. London, Times Newspapers Limited. Gale Document Number CS251994359.

[63] Plug, "The origin and early history of S_2A_3", p. 5.

[64] Hofmeyr, "Africa and Science", p. 273.

[65] P.R. Warhurst, "Smuts and Africa: a study in sub-imperialism" in *South African Historical Journal*, 16:1, 1984, pp. 82–83. http://www.tandfonline.com/doi/pdf/10.1080/02582478408671588, Accessed 5 December 2016.

associated with science was the notion of progress and civilisation. *The Times* of London portrayed "science and civilization [as] creeping into the Dark Continent".[66]

Clearly evident was a lack of self-reflexivity regarding the relationship between science and progress—and, with it, "civilisation". In the wake of the 1929 visit, a meeting was held by the British Association and the Royal Empire Society and, on the agenda, was the role of science in the empire. Lord Lloyd spoke of the importance of the role of engineering in obviating the periodic famines that had racked the Indian subcontinent as well as the better quality of life afforded by scientific endeavour in the Sudan. Sir Richard Gregory focused instead on the recent visit to South Africa. Improvement in the more distant parts of the Empire like South Africa "where the population was largely illiterate" was best fostered by state intervention acting in conjunction with scientific experts. Science had an integral role to play in inculcating the values of modernity and in addressing the "native question". Through anthropology, indigenous custom could be understood so as to aid administration. South African agriculture also relied on "cheap native labour" and the modernisation of agriculture would be necessary should "the native problem become more acute" leading to a less compliant workforce.[67]

Jan Hofmeyr's speech to the British Association in 1929 alluded then both to South Africa as the heir to a European tradition where notions of science—as derived from the Enlightenment—were synonymous with both nationalism as well as imperialism. This was reinforced by the continent of Africa itself and its increasingly important position in natural history as the birthplace of modern humans but also as the remnant of a prehistoric supercontinent that had rendered national boundaries superfluous: "This great land-mass which has reared itself against time's passage, almost since time's beginning".[68]

On the 4th of August in Johannesburg, a meeting was held that comprised the zoology, botany and geology contingent of the British Association. The subject of discussion was Gondwana. The head of the Zoology Section, Professor Watson, focused on the common fauna,

[66] "Science in Africa" in *The Times*, July 23, 1929.

[67] "Science and the Empire: Sir R. Gregory on Research" in *The Times*, 4 December 1929, Issue 45377, p. 16. London, Times Newspapers Limited. Gale Document Number CS268902788.

[68] Hofmeyr, "Africa and Science", p. 274.

particularly the similarity between reptiles in Africa and South America. He also cited the example of the lungfish—a unique fish capable of surviving on land—that could only be found in South America and South Africa.[69]

In Alex Du Toit's presentation, he asserted that Gondwana's existence was "almost indisputable". The only real area of contention was whether the clearly observed correlations in terms of rock, fossil and zoological evidence was due to the connection of continents by "land-bridges" or that these continents had, in fact, once been joined together in a vast landmass. Du Toit was clearly in favour of the latter and illustrated his presentation by drawing a map for the attendees where he suggested the possible early position of these continents, substantiating his points with geological evidence (including that of *Glossopteris*).[70]

Du Toit's presentation, however, was not as indisputable as he had hoped and he was met with scepticism. A representative from the Australian Association for the Advancement of Science, Dr Walkom, believed that the common presence of only one type of flora was insufficient to support such an ambitious hypothesis. He was supported by the University of Vienna's Professor Abel who emphasised that far more evidence was required before present conceptions could be adapted to allow for a hypothetical supercontinent.[71]

Du Toit's undoubted frustration at the incredulous responses was somewhat ameliorated six years later. At a meeting of the British Association held in Norwich in September 1935, the presidential address was delivered by Professor W.W. Watts and entitled "Form, Drift, and Rhythm of the Continents", a *volte-face* from the hostility of 1929. Part of Watts's speech focused on the similar rock formations and fossil material evident in the northern hemisphere positing a connection that he termed "the Palaearctic Region". He cited Du Toit's work extensively in conjunction with the earlier conclusions made by Wegener, highlighting the former's use of glacial evidence and the observed stratigraphic correlations made by Du Toit during his trip to South America. Discussed, too, was Du Toit's map which reconstructed the hypothetical Gondwana. While acknowledging that continental drift was far from universally accepted, Watts concluded with a sterling expression of support:

[69] "Discussion on a Lost Continent: Evidence from Fish" in *The Times*, August 5, 1929, Issue 45273, p. 9. London, Times Newspapers Limited. Gale Document Number CS15272869.

[70] "Discussion on a Lost Continent: Evidence from Fish" in *The Times*, August 5, 1929.

[71] "Discussion on a Lost Continent: Evidence from Fish" in *The Times*, August 5, 1929.

> Thus, when all is said and done, movements on a colossal scale are established facts, and the question of the future is how far we shall accept the scheme of drift due to Wegener, or one or other of the modifications of it. It is for us to watch and test all the data under our own observations, feeling sure that we shall have to adapt to our own case Galileo's words, *e pur si muove*.[72]

While the visit of the British Association in 1929 stimulated the enthusiasm of scientists (and the public) in general, yet another conference was held that same year which catered specifically to geologists and was another marker of South African assertion of nationhood within the international scientific community.

The International Geological Congress

The first International Geological Congress (IGC) was held in Paris in 1878,[73] the year of Alex Du Toit's birth; however, it was conceived almost two decades earlier. It was the brainchild of Giovanni Capellini, an Italian geologist, who was particularly skilled at geological mapping and, like Du Toit, represented the scientific endeavour of a newly emergent nation. For much of Capellini's career, he travelled extensively, both in Europe and in North America, corresponding with geologists from all over the northern hemisphere. He had come to realise that the Earth's geology knew no national boundaries and that it was essential for the development of the science for open communication and the sharing of information that would allow for the comparisons so essential to stratigraphy. His trip to North America lasted almost five months and comprised a geological exploration of Canada and the United States, where he studied the geology of local areas, collected rocks and fossils and engaged with prominent geologists, some of whom would go on to initiate the IGC. It has been surmised that it was during this trip that Capellini suggested the creation of an organisation of geologists that would exchange information and reach international consensus in matters of stratigraphy and related fields. In a letter to H.E.G. Finali, the Minister of Agriculture, Industry and Trade in Italy, in early 1874, Capellini proposed:

[72] "The British Association Opening of Norwich Meeting: Professor Watts on the Chain of Life" in *The Times*, September 5, 1935, Issue 47161, p. 6. London, Times Newspapers Limited. Gale Document Number CS100871461.

[73] "The International Geological Congress (A Brief History) http://iugs.org/uploads/images/PDF/A%20Brief%20History.pdf, Accessed 2 November 2016.

> the opportunity to seek first an agreement among the geologists of different nations about nomenclature and systematic subdivision of the different terrains, and to invite the Directors of the Geological Institute and the most distinguished geologists of the different nations to an International Congress to be held in Rome.

It was, however, at a meeting for the American Association for the Advancement of Science two years later, that the IGC was established. Many of Capellini's friends and colleagues would become part of the founding committee and Capellini was himself made Vice President of the inaugural IGC held, not in Italy, but in Paris in 1878.[74]

Unsurprisingly, this first IGC was dominated by Europe and North America, with the exception of Germany which was still harbouring ill-feeling towards France in the wake of the Franco-Prussian War. Like its latest incarnation in South Africa, the IGC took place from late August to early September but was substantially smaller with 312 officially registered delegates (the majority of whom were from France) and the presentation of 41 papers. The subject matter of the IGC would have been familiar to the modern geologist—the "Standardization of geologic maps and reports with regard to nomenclature and symbols", a consideration of rock systems and faults and, what would be of particular importance to continental drift theory—the use of animal and plant fossils in determining rock systems and the means of discerning rock "origin and age". This IGC was not without its shortcomings—its representation was limited, the papers were eclectic and it was plagued by the perennial problem of all conferences—the time allocated for the presentation of papers was entirely too short (15 minutes).[75] It was, however, at the 2nd IGC that real strides were made in terms of international co-operation.

Capellini's role in initiating the IGC was recognised when there was unequivocal agreement to the proposal that the 2nd IGC be held in Bologna, Italy with Capellini as president. Taking place in 1881, it was here that formal steps were taken to establish international co-operation in geology. Also, beginning in late September, the programme of the IGC

[74] Gian Battista Vai "Giovanni Capellini and the origin of the International Geological Congress" in *Episodes*, Vol. 25, No. 4, December 2002. http://iugs.org/uploads/images/PDF/Capelllini.pdf, Accessed 2 November 2016, pp. 248–249, 249, 250, 252.

[75] Francois Ellenberger, "The First International Geological Congress (1878)" http://iugs.org/uploads/images/PDF/1st%20IGC.pdf, Accessed 2 November 2016.

made special provision for the adoption of geology nomenclature and symbols that would be internationally accepted and recognised: the rules according to which fossils and minerals would be named as well as the adoption of names for stratigraphic layers. Those who contributed were the experts in their field, including Eduard Suess. The discussions were intense with terms such as "formation", "terrain" and "system" provoking heated debate amongst the participants. Moreover, participants were divided over representing international consensus and co-operation and representing national concerns. This fell, of course, within the period of New Imperialism and concerns about prestige related to the use of *their* geological nomenclature and methods was at the forefront of the nationalist contingent. Capellini represented the international contingent and did his best to ameliorate these tensions, ensuring the co-operation of the major powers, until agreement was reached. However, the IGC was itself largely composed of members from the northern hemisphere and the final adoptions thus reflected the dominance of this part of the world, related to their economic and military dominance in the late nineteenth century. Nevertheless, the spirit of hard-won co-operation between rival powers hinted at future scientific collaboration and members were optimistic that:

> they could well reach beyond the problems of finding a common international geological language and seek comparable reconciliation on key issues in pure science of common interest.[76]

Of particular interest here is the way in which international rivalries played out within the context of scientific discussion with tension existing, for instance, between German and French delegates.[77] Geology was not neutral terrain. The agreements reached, however, allowed for the fostering of international co-operation and gave geologists the ability to make connections across national boundaries related to stratigraphy and palaeontology. This would be essential to the work of continental drift theorists. However, reflective of the balance of power, the subsequent location of IGC meetings mirrored the dominance of North America and Europe until 1906 when the 10th IGC was held in Mexico City. The first held in the southern hemisphere occurred in Pretoria, South Africa in 1929.[78]

[76] Gian Battista Vai. "The Second International Geological Congress, Bologna, 1881" in *Episodes* Vol. 27, No. 1. http://iugs.org/uploads/images/PDF/2nd%20IGC.pdf, Accessed 2 November 2016, pp. 14, 16–18.

[77] Vai, "The Second International Geological Congress", p. 17.

[78] "The International Geological Congress (A Brief History)"

The decision to host the Congress in South Africa was somewhat anti-climactic. With no chosen venue for the 15th Congress as the 14th Congress concluded in Madrid, Spain, the president of the IGC, Don Cesar Rubio, wrote a letter to A.L. Hall, the South African delegate at the Congress, raising the possibility of holding the following Congress in South Africa due to the country's mineral resources and mining industry.[79] Based on Rubio's request to obtain permission from the South African government to hold the following IGC in the country, Hall—with the support of the Secretary for Mines and Industries, H. Warington Smythe, and the Acting Minister for Mines and Industries, F.H.P. Cresswell—was able to obtain financial aid from the Union Government for the 15th IGC.[80] With further financial assistance from mining concerns, the Geological Society of South Africa and the cities of Kimberley, Cape Town and Pretoria and with the belief that it would be opportune to hold the IGC at the same time as the visit of the British Association for the Advancement of Science, a message was sent to Madrid: "Prime Minister agrees on behalf of Union Government invite Congress meet here 1929 probably August. Cable whether you accept".[81]

With the subsequent issuing of a formal invitation by the South African government, preparations for the 15th IGC got underway. At a specially convened meeting of the Geological Society of South Africa, Du Toit—then the president of the GSSA—addressed the members, pointing out the nomadic nature of the IGC and its importance in bringing together geologists from all over the world with the aim of sharing knowledge and gaining some familiarity with each country's unique geology. For Du Toit, not only did the IGC present an opportunity for South African geologists to make international connections, it allowed international geologists to be exposed to South African geology and his words echoed the distinctive relationship between the individual and the global that defined much of his career:

[79] Council for Geoscience—Library Information Centre: Letter by Cesar Rubio to A.L. Hall, 28 March, 1926 reprinted in *IGC*, Vol. 25, No. 1, 1929, p. 4.

[80] Council for Geoscience—Library Information Centre: *IGC*, Vol. 25, No. 1, p. 4.

[81] *IGC*, pp. 5–6.

> To be able to meet these eminent specialists with an international repute to discuss our peculiar problems with them: to draw upon their extensive and varied knowledge and finally to show them in return some of our novel and, in some instances, unrivalled sections and mineral deposits, would indeed be a mutual education and lead to the advancement of geological thought the world over and in South Africa in particular.[82]

The Organising Committee of the 15th IGC was elected at this meeting and included Prime Minister Hertzog as the Honorary President with A.W. Rogers as President. Du Toit later joined the Organising Committee in 1928. The IGC was to be based at the Extra Mural Buildings of the Transvaal University College—the current University of Pretoria. In homage to the mineral resources which had gained South Africa international repute and come to define the country's economy and history, the theme of the 1929 IGC was "The Gold Resources of the World".[83]

Along with intellectual considerations were the logistics. This was compounded by the IGC's requirement of having excursions to various geological features of interest. These excursions began even before the conference officially got under way and continued past its official close and ranged from the scenic spots of Cape Town—Chapman's Peak, Table Mountain and Sea Point—to various mines in Johannesburg, the Vredefort Granite Dome and sites as far afield as Rhodesia. The recommended book to help familiarise them with South African geology was Du Toit's *The Geology of South Africa* priced at £1.8.0. The involvement of both the Union and Rhodesian governments was apparent in terms of a travel concession allowing IGC members a discounted fair on railway travel and efforts were made to alleviate the high cost of steamship travel from the northern hemisphere, highlighting South Africa's distance from the metropole. To further ease the transition between northern and southern hemispheres, delegates were reminded that it was winter in South Africa and that tropical attire would not be suitable.[84]

For Du Toit the International Geological Congress that was subsequently held in Pretoria in 1929, almost in conjunction with the meeting

[82] Council for Geoscience—Library Information Centre: Address by Alex Du Toit to Special General meeting of the GSSA, 27 April, 1927, reprinted in *IGC*, Vol. 25, No. 1, 1929, p. 9.

[83] *IGC*, pp. 10, 23–24, 13.

[84] *IGC*, pp. 31–41, 44, 42.

of the British Association in Johannesburg, marked the peak of geology in South Africa: "Never had such a galaxy of talent become aggregated in this country".[85] Advantage was also taken of the proximity of the two visits when the British Association invited the IGC members to the presidential address by Sir Thomas Holland in Johannesburg.[86] The Organising Secretary of the IGC committee in South Africa was A.L. Hall and its president was Du Toit's colleague and mentor, A.W. Rogers. Preparations had begun for the IGC a year earlier with the creation of programmes and itineraries for excursions. Also important were the guidebooks penned for the occasion with Du Toit as one of the authors.[87]

There were 22 guidebooks dealing with significant aspects of southern African geology from Cape Town to Northern Rhodesia which had been written by prominent geologists of the period. They were designed to accompany the visiting geologists as they undertook excursions around the country. A.W. Rogers wrote an introductory guide to the geology of South Africa—based on the early work that he had undertaken with Du Toit—as well as the geology of Johannesburg. S.H. Haughton focused on the eastern and western Cape. For his part, Du Toit co-authored the guidebooks relating to the Cape and Natal.[88] His description of the geology of Kimberley focused on diamond mining with a detailed description of the diamond-bearing kimberlite.[89] He was also able to draw upon his expertise in glaciation (which had initially substantiated his views on drift) in his discussion of the pattern and direction of glacial movement evident in the rocks underlying the Karoo System.[90] The glaciation that characterised the Carboniferous Period was also covered in his description of the Natal midlands between Durban and Pietermaritzburg with the visible evidence of glaciation in the form of tillite and erratics. And it was beneath the sandstone of the Lower Beaufort Beds in Northern Natal that

[85] Alex L. Du Toit, "Arthur William Rogers" in *South African Journal of Geology*, Vol. 49, pp. 291–304, 1946. http://www.sabinet.co.za. AJA10120750_1936, Accessed 13 October 2016. p. 295.

[86] *IGC*, p. 67.

[87] Du Toit, "Arthur William Rogers", p. 295.

[88] Geological Society of South Africa (GSSA): "International Geological Congress XV Session, South Africa, 1929: Guidebook" (Pretoriam Wallachs Ltd, 1929).

[89] GSSA: S.H. Haughton and A.L. Du Toit "Cape-Kimberley"—Excursion A.5, pp. 19–20.

[90] GSSA: A.L. Du Toit and A.W. Rogers "Cape-Kimberley"—Excursion A.6, pp. 8–9.

Glossopteris fossils had been found.[91] The guidebooks were merely descriptive with no reference to continental drift. Yet this was the work most associated with Du Toit at the IGC, giving him a certain notoriety, as evident by Alpheus F. Williams, a representative of De Beers: "'We must associate everything with the name of Dr Du Toit,' whose work was well known; in fact, 'Dr Du Toit's book had been under every visitor's arm'".[92] Du Toit's diary entries for the three weeks covering late July and early August demonstrate the punishing schedule required to host both the British Association and the IGC:

- Monday, 22nd July—"Busy again…Took party I.G.C. to Riverton Saltpan in morning. Luncheon at home. Saw scientists off by 5.45 train"—met successive groups of I.G.C.
- Followed closely by arrival of British Association—Monday, 29th July—B. Assn main party arrived"; 30th July—"Very busy. BA Assn second party"
- Wednesday, 31st July—"Arrived at Pretoria and attended I.G.C. Congress"
- Tuesday, 6th August—"Meetings of I.G.C.—busy"[93]

In addition Du Toit presented papers at the IGC that reflected his various geological interests which underpinned his belief in continental drift, including "A Brief Review of the Dwyka Glaciation of South Africa" and "A Short Review of the Karoo Fossil Flora".[94]

The strength of the IGC was its drawing together of geologists from all over the world, allowing for the reconstruction of past Earth processes—particularly relevant to continental drift theory—across

[91] GSSA: A.L. Du Toit and E.C.N. van Hoepen "Cape-Kimberley"—Excursion C.18, pp. 4–7.

[92] UCT-JL: Alex L. Du Toit Papers—BC 722: A4—Personal Documents. Newspaper clipping, 26 July 1929.

[93] UCT-JL: Alex L. Du Toit Papers—BC 722: A1—Diaries: 1929.

[94] Council for Geoscience—Library Information Centre: *International Geological Congress: Compte Rendu of the XV Session, South Africa, 1929—Vol II: Scientific Communication*. (Pretoria, Wallachs' Ltd, 1930) piv.

national boundaries. Despite the success of the event, Du Toit was disappointed that he was unable to interact with South American geologists who had not attended and whose presence—and knowledge of South American geology—would have bolstered the claims of mobilists. It would have offered further opportunity to continue the comparisons made between the two continents that Du Toit had begun in 1923:

> I had confidently hoped that during the visit of the International Geological Congress some geologists would have come over from S. America, but unfortunately that great & important continent was wholly unrepresented! It was indeed to me an immense disappointment, for I had expected to have had the great privilege of showing some of our outstanding problems to S. American colleagues.[95]

However, the IGC had been beneficial to South African geology as well. In a prelude to the 16th IGC that was to be held in Washington in 1933, Du Toit assessed the impact of the IGC on South Africa. First and foremost, it had served as a means of convincing the large proportion of visiting geologists both of the unique nature of South African geology and of the explanations for this developed by South African geologists. Some delegates were, simultaneously, able to take advantage of their time in South Africa to carry out limited research and others disseminated their knowledge through various publications and in a multitude of tongues. The repercussions were concrete—C. Freire De Andrade and A.J. de Freitas encouraged the Portuguese to carry out a geological survey of their colony in Mozambique. In addition, a "Sub-commission of the African Surveys" was created that subsequently published the Geological Map of South Equatorial Africa. For Du Toit, this focus of scientists "[marked] the beginning of a new epoch in the geological investigation of the 'Dark Continent'". This was in contrast to the possibilities of the following IGC in the United States which, for Du Toit, was a territory already comprehensively explored and mapped by geologists and offered little of the excitement of revealing the geological complexity of the "Dark Continent".[96]

[95] UCT-JL: Alex L. Du Toit Papers—BC 722. B2—Correspondence with colleagues in South America (ca. 1920s–1931): Letter by Du Toit to Sr E. Terra Arocena Ingeniero-Director, Instituto de Geologia y Perforaciones, Montevideo, Uruguay, 17 December 1929.

[96] Council for Geoscience—Library Information Centre: B0301693—Alex Du Toit: "The Aftermath of the Fifteenth International Geological Congress: An appreciation from South Africa" in *Economic Geology*, Vol. 28, pp. 389–391.

Du Toit's assessment was a clear exercise in the way in which the local interacted with the global and was an assertion of South African identity even as it acknowledged the dominance of the northern hemisphere.

Conclusion

It was a triumph for South African science on 23 September, 1931 when Jan Smuts was named president of the British Association for the Advancement of Science at a meeting held in Westminster, London. From a British perspective, Smuts's investiture was viewed as the culmination of a long process that had begun with a narrow focus on science within the confines of the United Kingdom. With the expansion of empire, however, and the growing involvement of the colonies and, later, Dominions, science had an important role to play in the transmission of British values and ideas of progress, civilisation and modernity. Smuts emphasised his own academic background as the product of a British university and his inaugural address was imbued with the imagery of empire:

> The Association...had during the century now passed run a course not unlike that of the Empire on a smaller scale. It had become the parent of similar science associations in most of the Dominions, and this centenary meeting had truly become a great family reunion.[97]

Smuts concurred with the optimistic view of science as having the potential to solve all the ills that beset humankind and his address highlighted the significant events in the history of science that had characterised the past one hundred years of the Association's existence. As Smuts concluded, however, he emphasised that the transmission of scientific knowledge had not been a unidirectional process. The British Association visits to parts of the Empire had introduced British scientists and intellectuals to the unique nature of the Dominions—their resources and challenges—and this, in turn, had shaped their perceptions and broadened their understanding. While the characteristics of the scientific method and the production of modern scientific knowledge may have largely been credited to European enterprise, it was no longer confined to Europe.[98] The child had become an adult.

[97] "The British Association: London Meeting Opened" in *The Times*, 24 September 1929, Issue 45936, p. 17. London, Times Newspapers Limited. Gale Document Number CS117646648.

[98] "The British Association: London Meeting Opened" in *The Times*, 24 September 1929.

CHAPTER 9

Our Wandering Continents: Du Toit's Definitive Work, Controversy and Consensus

With the publication of *Our Wandering Continents* in 1937, Du Toit was able to create a more accurate representation of Gondwana,[1] which is easily rendered by computer models today based on satellite data capable of measuring the direction and rate of plate movement. Du Toit, however, had no access to this and *Our Wandering Continents* represented the culmination of his extensive field research.

Our Wandering Continents could not, by any means, be defined as tentative. Including the indices, it ran to 366 pages and represented both a culmination and a synthesis of Du Toit's work on continental drift. Dedicated to Alfred Wegener, Du Toit's aim in its publication was to lend geological weight to Wegener's hypothesis.[2] The dedication adopted an almost martial tone with Wegener acknowledged for his "distinguished services in connection with the geological interpretation of our earth"—perhaps a subconscious acknowledgement of the controversy and opposition that the work would arouse, the hardening of battle lines.[3] It is,

[1] Philip Stone, "Geological exploration of South Atlantic islands and its contributions to the continental drift debate of the early 20th century," in *Proceedings of the Geologists' Association*, 2015, Vol. 126, 266–281, p. 12. Taken from http://nora.nerc.ac.uk/510724/1/SouthAtlanticIslands-pga.pdf. Accessed 12 October 2016.

[2] Robert Muir Wood, *The Dark Side of the Earth* (London, George Allen and Unwin Publishers Ltd., 1985), p. 107.

[3] Alex L. Du Toit, *Our Wandering Continents: An Hypothesis of Continental Drifting* (London and Edinburgh, Oliver and Boyd, 1937).

S. Chetty, *'Africa Forms the Key'*, Cambridge Imperial and Post-Colonial Studies, https://doi.org/10.1007/978-3-030-52711-2_9

however, the title page that is of particular interest. It contains an isolated phrase in parenthesis, "Africa forms the Key".[4] There are two possible interpretations of this phrase—the first is that it is spatial. Africa was a central part of Gondwanaland, its rocks the evidence for both the existence of the supercontinent and of continental drift. The other related significance of the phrase is that it is ideological. Asserting the significance of the African continent in the history of geology and human evolution was central to an avowal of South African identity as an independent southern nation. It moved the continent—and the country—from the periphery to the centre, both literally and figuratively. Shaped by the European Enlightenment, South African science nevertheless played its own unique and pivotal role, making a significant contribution to the history of the world.

Drawing upon the history of science, Du Toit envisioned continental drift to have as profound consequences as Darwin's theory of evolution. This "New Geology"—as he termed it—would have implications for fields ranging from studies of climate to geography yet it faced an almost insurmountable obstacle—the "psychology" of the disbelievers who would not be persuaded by its objective validity. Simultaneously, Du Toit was himself painfully aware that drift could not be proven and still required an understanding of the physics that underlay the movement of the continents—which would form the core of the criticism aimed at the book.[5]

The Preface begins by demonstrating that Du Toit was unable to give a suitable description of the more than 250 million-year-old fold systems of the southern hemisphere that did not take into account continental drift—for him, drift had become an integral means of interpreting geological formations. This opening paragraph also less than subtly criticises prevailing thought that explained little, "things are *there* because they *are* there" [Du Toit's original emphasis]. Further criticism was aimed at those who had not for themselves assessed the feasibility of the hypotheses put forward by Taylor and Wegener before dismissing them. In contrast, for Du Toit, drift had become the "revolutionary Hypothesis" of "truth" which encompassed the observable geology of the world around, both explaining and predicting geological processes. His use of "revolutionary" was particularly apt—continental drift was envisioned as nothing less than a

[4] Du Toit, *Our Wandering Continents*.
[5] Wood, *The Dark Side of the Earth*, p. 108.

profound rethinking of current geological understanding, a true revolution in the earth sciences.[6] As prefaces go, it was ambitious.

Our Wandering Continents is composed of 17 chapters, framed by an opening chapter which assessed continental drift in relation to other contemporary theories and a final chapter which considered the possible mechanisms for continental drift. The opening chapter continues in the spirit of the Preface, placing continental drift in opposition to what Du Toit termed "orthodoxy" and a "conservatism" or static approach to geology that was both accepting of conventional understandings—despite possible shortcomings—and equally hostile to new ideas. He pointed out the lack of consistency in "current theory" that was undecided with regards to the origin of the rift valley, to the strength of the crust, the fluidity of the mantle and the width of land bridges. The only source of agreement apparently was their rejection of drift. Criticism was also reserved for Lyell's "cramping influence of Uniformitarian doctrine" that could be interpreted to argue for fixed continents and ocean basins. Yet, Du Toit argued, it was the very postulation of the existence—and subsequent disappearance—of land bridges that contradicted the notion of permanent "deep oceans". He went on to list the various flaws in contemporary understandings of the Earth, culminating in the failure of the notion of fixed continents to account for the stratigraphic, biological and climatological convergence across continents. Ultimately, while continental drift was not itself without flaw, the same could be said for its orthodox rivals.[7]

The subsequent chapter dealt with the history of continental drift theory. Tracing allusions by eminent Enlightenment figures such as Francis Bacon and the Comte de Buffon to continental drift as far back as three hundred years previously, Du Toit was able to establish an illustrious pedigree for his work. In the twentieth century, influential figures advocating drift were Frank Bursley Taylor who was published in 1910, Herbert Baker and Alfred Wegener published in 1912.[8] Baker had been the Government Geologist for the Falkland Islands and had noted similarities between the Table Mountain Series at the Cape and the rock formations of the Falklands, with drift being the most likely explanation for his

[6] Du Toit, *Our Wandering Continents*, p. vii.

[7] Du Toit, *Our Wandering Continents*, pp. 1, 2, 7–8, 9–10.

[8] Du Toit, *Our Wandering Continents*, p. 11.

observations.[9] The remainder of the chapter details contributions to drift theory by a number of other figures and this exhaustive itemisation is summarised by Du Toit in the conclusion. Drawing together these disparate voices was no easy task but Du Toit was able to find several common points of agreement—among them was the consensus that two landmasses had existed in the Palaeozoic era—Gondwana and Laurasia—and that these had subsequently begun to break apart towards the latter part of the Mesozoic era, a movement that was still in progress. Ultimately, these theorists were of the opinion that drift was a mechanism that had occurred throughout the Earth's geological history and that its origins lay within the interior of the Earth.[10]

Du Toit was all too aware of the uphill battle he faced in Chapter III. Entitled "Geological Principles", the chapter focused on the various components of the Earth's surface and interior that formed part of the mechanisms of drift. Metaphorically, however, it was akin to building a house on an unstable foundation as Du Toit admits in his introductory sentence:

> Every "Hypothesis of Continental Displacement" must of necessity have as its basis the known or inferred properties of the crust or lithosphere of this unstable Earth of ours, though unfortunately such forthwith involves the very branch of geological science wherein opinion is most diverse and in many respects least convincing.[11]

It was, as has previously been mentioned, the explanations for the mechanism of drift that were most subject to contestation. Much of the activity of the Earth's interior had been conjecture that would only be supported by the new technology that still lay a decade into the future. In his attempt to explain drift, however, Du Toit postulated the importance of isostasy.[12]

The principle of isostasy has a longer history than contemporary understanding of the division of the mantle into the more rigid lithosphere which lies atop the weaker asthenosphere, yet has been incorporated into current geological understanding. With its origins in the eighteenth century, where the gravitational attraction exerted by the Andes mountain

[9] Henry R. Frankel, *The Continental Drift Controversy: Volume I: Wegener and the Early Debate* (Cambridge, Cambridge University Press, 2012), p. 455.

[10] Du Toit, *Our Wandering Continents*, p. 36.

[11] Du Toit, *Our Wandering Continents*, p. 37.

[12] Du Toit, *Our Wandering Continents*, pp. 42–43.

range was found to be smaller than expected, French surveyors accounted for this anomaly by proposing that the mountain range contained a "root" that extended into the mantle which acted as a counter-balance to the weight of the mountain range. Isostasy exists where relatively rigid material (in this case, the crust or lithosphere) rests on a weaker, more "ductile" layer (the asthenosphere). The rigid crust moves vertically (either in an upwards or downwards direction) in order to reach an equilibrium. This can be illustrated through the erosion of mountain ranges such as that of the Andes—erosion will decrease the weight exerted by the mountain range and it will thus rise to compensate with its gravitation attraction decreasing. The sediment derived from the erosion will accumulate in basins which will subsequently sink as a result of the additional weight. The same applies to glaciation where the weight of the ice sheet will lead to a sinking of the crust and the corresponding movement of the asthenosphere to compensate for the increased weight while the retreat of the glacier will lead to the crust rising due to the reduced weight load.[13] For Du Toit, then, isostasy became a key mechanism in explaining the movement of the asthenosphere and, with it, the lithosphere or crust.

The penultimate section of the chapter "Criteria for Continental Drift" deals with the type of evidence that may be used to suggest drift and Du Toit is careful to point out that it is less about isolated moments of similarity than it is about a preponderance of evidence. The criteria include the "Physiographical" evident in the ostensible fit between separated coastlines; "Stratigraphical" relating to a continuation of rock formations and layers across continents that may be correlated by rock type and fossils; "Tectonic" which deals with the folds, faults and rift valleys that are continuous across distance and, in some cases, ocean; "Volcanic" showing the simultaneous occurrence of past eruptions; "Palaeontological" based on the flora and fauna that form part of the fossil record and "Geodetic", the actual measurement of latitude and longitude to determine changes as a result of drift. In subsequent chapters, Du Toit then proceeded to apply these principles in his reconstruction of Gondwana and, in Chapter VI, drew upon his own research in South America and South Africa a decade earlier to apply the criteria for drift. Drawn into this discussion were the other landmasses that had once comprised the supercontinent—the Falkland Islands, Antarctica, Australia, Madagascar and India. More

[13] *The Oxford Companion to the Earth*, P.L. Hancock and B.J. Skinner (eds.). (Oxford, Oxford University Press, 2000), pp. 559–569.

ambitiously, Chapter VII saw Du Toit apply these criteria north of the equator to Laurasia emphasising that, as with Gondwana, "the polar region provides the key to the elucidation of Laurasia". The difference between the two supercontinents, however, was that Gondwana had been a single large mass and thus only subject to distortion at its edges whereas Du Toit termed Laurasia "a composite mass" composed of the accretion of numerous bits of crust that created a haphazard arrangement of deformation making it a challenge to reconstruct (or deconstruct). He nonetheless perseveres, tracing connections between Newfoundland in Canada with Scandinavia, Britain and Ireland. The evidence employed in this chapter was based on stratigraphy and also largely related to orogeny and the tracing of mountain chains across continents with the subsequent chapter emphasising the use of fossil evidence and Du Toit's hypothesis that Laurasia had at one point been composed of a minimum of six "sub-continents". Chapter VII concludes with an encouragement to the reader to familiarise themselves with the relevant work of Dutch geologist, Waterschoot van der Gracht (a significant European proponent of drift who is discussed later in this chapter).[14]

With an almost biblical resonance, Chapters IX and X describe the history of the land and that of the oceans respectively. In similar vein, the introduction detailing the history of the landmasses suggests that the Tertiary period was marked by cataclysm as continents ruptured and came together in new ways causing further deformation. The result was "two closed orogenic rings"—one surrounding Gondwana and the other Laurasia. Accompanying maps and diagrams—first looking at the Earth as a whole and then at particular regions—demonstrate surreal reality with the familiar continental shapes placed in eccentric positions, on their way to becoming the global map with which humankind is familiar.[15] Accompanying the visual reconstructions are detail-heavy descriptions that prove difficult to absorb. What they point to, however, is Du Toit's ability to amass extensive research on a global scale and use this to reconstruct a past world.

In the subsequent chapter, Du Toit turned his attention to the formation of the oceans, disputing two existing theories. The first was that most associated with American geology and considered that the ocean basins

[14] Du Toit, *Our Wandering Continents*, pp. 51–53, 106–132, 133, 137, Chapters VII and VIII, pp. 158, 157.

[15] Du Toit, *Our Wandering Continents*, pp. 174–177, 180.

were a perpetual fixture, immune from change. The second hypothesised the elevation and sinking of crustal blocks leading to the position of the oceans shifting accordingly from higher to lower ground. Neither harmonised with Du Toit's view of drift and both were unceremoniously discarded. The oceans owed their origin to a single mother ocean "Panthalassa" and it was through the horizontal movement of crust that the ocean became subdivided, concentrating in the areas where they are found today. This was in contrast to the vertical displacement hypothesis which Du Toit termed "arbitrary". Based on the facts that the oceans lie at great depth while land is at higher elevation, he used the principle of isostasy to suggest that the underlying oceanic rock was largely Sima or basalt, a denser material than the largely granitic composition of crust hence the greater difference in elevation between land and sea.[16] An account of the formation of the Pacific, Atlantic and, particularly, Arctic Oceans is rather summarily concluded with an unequivocal dismissal of the land bridge hypothesis with which the chapter ends. The use of arbitrary land bridges to explain animal migrations across continents now separated by ocean is considered a poor substitute for displacement:

> The Displacement Hypothesis is ... competent to explain these and other puzzles of biological distribution ... in a simple and logical manner and without violation of isostatic principles. **Current views of Continental Linking must therefore be firmly rejected**. [Emphasis in original][17]

Much of the later chapters of *Our Wandering Continents* deal with the evidence for continental drift, evidence that ranged from the hypothetical to observations for which drift is posited as the only solution. Du Toit devotes two chapters to the "Paramorphic Zone" which, in conjunction with isostasy provides the mechanism for driving drift from the Earth's interior. To emphasise the importance of the paramorphic zone, he builds an argument that begins with the basic understanding of the structure of the Earth and its relation to isostasy. The sial or the topmost layer of the Earth's crust lies upon the sima, a lower and "weaker" layer which both exist in "isostatic equilibrium". Any addition to the sima such as sedimentation or decrease such as erosion impacts the equilibrium and the matter upon which it lies flows from one area to another in order to compensate

[16] Du Toit, *Our Wandering Continents*, pp. 210, 212.
[17] Du Toit, *Our Wandering Continents*, p. 228.

for the change in density and restore equilibrium. This, then, provides a plausible mechanism for drift which is further illustrated with Du Toit's use of the paramorphic zone, a region within the Earth's interior of high density and pressure as evident from the metamorphic minerals formed here such as garnet, diamond and zircon (with the last later becoming an important means of reconstructing plate tectonic movement). For Du Toit, then, it is clear that the zones of differential pressure further affect the isostatic equilibrium and contribute to drift. Related to this is the notion of heat which may contribute to the expansion of the crust and offset the contracting effect of gravity although, here, Du Toit admits that more research needs to be done.[18] Drawing upon existing understandings, his explanation was plausible but lacked scientific rigour in that his hypothesis was based on an unknown region and failed to fulfil the criteria of testability.

His subsequent chapter, however, attempts to compensate for this by moving the focus to the Earth's exterior and returning to the very beginnings of the discipline with James Hutton's attention to observable geological processes which were the foundation of uniformitarianism. Ironically, for drift detractors, drift theory and the rupturing and giddy move of continents seemed to smack of catastrophism yet, for Du Toit, this was a process that was ongoing and was simply uniformitarianism on a global scale. Like Hutton, Du Toit used the observable to infer processes that occurred in the inaccessible interior. River deltas were sites of sedimentation and glaciers and ice sheets similarly exerted tremendous weight on the crust which subsequently rebounded as glaciers retreated. Flat plains were the remnants of large-scale erosion promoting a decrease in density.[19] It was these changes in density that affected isostatic equilibrium: "The loading of a basin by sediment or by ocean brings about a related condensation of matter in depth; the unloading thereof a corresponding expansion", and it is this movement of "matter" as it seeks to restore equilibrium that Du Toit proposed carried with it the continents.[20]

The strength of what are increasingly dense chapters, however, lies again in Du Toit's ability to create a geological narrative. In this case, however, this was less about recreating the past than it was about predicting the future and therein lay the importance of drift theory—it was not

[18] Du Toit, *Our Wandering Continents*, pp. 229, 242, 230–231.

[19] Du Toit, *Our Wandering Continents*, pp. 243–248.

[20] Du Toit, *Our Wandering Continents*, p. 243.

just an explanation for existing geological features but a means of showing the future. This was evident in his discussion of rift valleys—areas where the crust is exposed to tensional forces, thinning the crust and creating a rift between the opposing forces, leading to an area of subsidence that may be filled with sediment. The thinned crust makes it prone to volcanic and tectonic activity. It is also the site at which continents break apart and, as Du Toit shows, Africa is the site where these processes can be observed. It is here that his language moves from dryly analytical to almost poetic:

> Obvious, indeed, is the "breaking-up of Africa." We observe the initial separation of Arabia; we see Madagascar long-separated and displaced to the south-south-east from its original position in the bight of Tanganyika; we perceive the curing sector from Somaliland to Beira that is about to follow suit; we notice behind it the Lake Victoria block not quite severed; and we detect in the hinterland, penetrating into the heart of the Continent, sundry fractures and extensive far-flung saggings of the peneplained surface. Can we doubt, then, that the partition of Africa from east to west is still in progress?[21]

In Chapter XIII, Du Toit moves away from geological evidence to focusing on palaeoclimate, highlighting the clear differences between past and present climates as evident in the stratigraphic record. These tremendous variations in climate included the evidence for extensive glaciation in the southern hemisphere on continents at present located at a distance from the polar region, milder climactic conditions in the northern hemisphere than is currently the case as evident by the ancient floral remains and marked dry conditions that prevailed on a global scale in the Triassic period. These anomalies in climate—at least from the perspective of the present—do not fit in with the current location of the continents in relation to the poles. Du Toit concurred with Wegener (a trained meteorologist) in observing that it was the disregard of continental drift that had created this predicament in understanding palaeoclimate. Building further on the work of Wegener, Du Toit elaborated on the features that represented different climactic zones such as glaciation, the formation of coal (derived from vegetation grown in both tropical and moderate climates—which was important for the Gondwana hypothesis), the salt and gypsum remnants of past deserts, the presence of coral limestone indicative of warm waters and plant and animal fossil evidence. His conclusion is

[21] Du Toit, *Our Wandering Continents*, p. 255.

simple—to account for these dramatic climatic changes between the past and the present, either the poles had moved in relation to the continents (which geophysicists had considered unlikely) or the position of the continents had changed in relation to the poles with a trend towards moving north from more southerly latitudes as evident with the Gondwana landmasses. For Du Toit, palaeoclimate then became "the most telling demonstration of the reality of Continental Drift".[22]

As in the case of climate, similar anomalies are evident in relation to fossil distribution—the focus of the subsequent chapter. Du Toit points out that the floral and faunal distribution as evident in the fossil record indicates "how out of accord existing terrestrial life is with present relationships".[23] His consideration of fossils meant that Du Toit returned once again to the prevailing notion of land bridges. These hypothesised connections between continents were breathtakingly varied: "bodies occupying the full length and breadth of the intervening ocean, through isthmuses, to mere chains of islands". Appearing as needed to account for the movement of plants and animals across continents, they quietly subsided into the ocean leaving no trace of their presence. There is understandably no little sense of sardonicism evident in Du Toit's discussion. Moreover, the principle of isostasy did not support the notion of sinking bridges while, ironically, they could be sunk if they were subject to forces of tension that could break them up. An even more compelling criticism of the theory was that the determination of their location was uncertain and arbitrary with little proof provided of their existence on the bridging continents. Summarily dismissing land bridges as "fundamentally unsound", Du Toit included within their ambit the romantic notions of lost lands such as Atlantis and Lemuria with the latter, ironically, considered to be part of the Gondwana hypothesis. However, just as Du Toit reserves the bulk of his criticism for the easily discounted land bridges, he does not sufficiently engage with the other plausible explanation of "parallel development or convergent evolution"—the similar evolution of unrelated species due to comparable environmental conditions and constraints. His somewhat lacklustre defence in this case is in the form of a circular argument—if the notion of continental drift is accepted based on other forms of supporting evidence then convergent evolution does not apply.[24]

[22] Du Toit, *Our Wandering Continents*, pp. 270, 272–274, 289.

[23] Du Toit, *Our Wandering Continents*, p. 290.

[24] Du Toit, *Our Wandering Continents*, pp. 290, 292, 294.

The final chapters of *Our Wandering Continents* deal briefly with geodesy and more extensively with orogeny. With regards to the former—and in line with Wegener's own attempts to measure drift—Du Toit is of the opinion that the process of triangulation would allow for the scientific measurement of drift. The existing means of measurement, however, were less than precise with earlier measurements suggesting a headlong rush of continents of as much as ten metres annually. The tracing of mountain chains, however, was something which had been central to Du Toit's comparisons with South America and South Africa and this was more familiar territory. A detailed discussion of the Earth's dominant mountain chains and corresponding deformation in the form of folding and compression brought him to the conclusion: "that the lands as a whole have pursued a strongly curved path over the face of the globe with anti-clockwise rotation".[25]

Du Toit concludes his work by acknowledging that much research is still to be done on continental drift; it remains nevertheless the strongest explanation for many of the puzzling observations for which biology, geology and climatology have been unable to account. Moreover, it is through an understanding of the process of drift and the accurate reconstruction of past continental movement that the theory will have its most important economic application—in the location of new mineral resources.[26] Simultaneously, despite its exhaustive detail and relentless exposition that traversed time, space and scientific discipline, much of what Du Toit proposed in *Our Wandering Continents* had been the subject of his attention since his earliest forays into continental drift. The work therefore represents both the synthesis and the culmination of his thinking on continental drift. And, just as the propositions put forward in *Our Wandering Continent* carried with them the air of familiarity, so too did the resulting reaction.

The reviews of the book ranged from the positive to the non-committal. Writing for *Discovery*, Malcolm Burr had clearly taken on board Du Toit's own criticism of those who opposed drift, highlighting the way in which continental drift theory had tapped into some deep-seated conservatism and opposition, both on the part of the general public and scientists in particular. It was this almost instinctual response—ranging from "apathy" to outright opposition—that made them immune to "the romance and

[25] Du Toit, *Our Wandering Continents*, pp. 299, 316.
[26] Du Toit, *Our Wandering Continents*, p. 332.

majesty of the conception". Despite, the theory finding favour with biologists in explaining faunal and floral similarity across continents, the same was not true for geologists and Burr had found it challenging to find copies of Wegener's seminal work that brought drift into the public eye. Attributing this lack of interest to a wariness regarding the "implications" of the theory—on which he does not elaborate—Burr nevertheless views it as an elegant means of providing a single overarching explanation for geological change.[27]

Other reviews, however, pointed to the explanatory mechanism of drift as being the key flaw in the theory. In *The Geological Review*, Du Toit's reliance on paramorphism was viewed as insufficient an explanation. This was accompanied by an unflattering description of the dense book with its unsuitability for the lay reader and necessary reliance on "tedious detail". Perhaps most jarring to the reviewer was Du Toit's tone which betrayed that of a "zealot advocate" rather than the idealised, neutrality of the scientist.[28] This was more subtly alluded to in a review that diplomatically considered Du Toit's openness to critique of various elements of drift theory due to his own belief in its truth: "Difficulties are generously admitted in the spirit of one who feels that his case is overwhelming".[29]

The lengthiest review of his book—as collected by Du Toit—was that penned by a South African colleague and the man who would write Du Toit's brief biography, T. W Gevers of the University of the Witwatersrand. Unconvinced by drift theory, Gevers' review is a masterful use of wordplay that details Du Toit's work on drift and his skills as a geologist while simultaneously distancing Gevers from an avowal of the controversial theory. At the same time, the subtitle of Gevers' review, "Brilliant Exposition by a South African" suggests a definite pride in Du Toit's work. Gevers begins by acknowledging Du Toit's formidable ability to construct coherence from an assemblage of observations and data which was put to good use

[27] UCT-JL: Alex L. Du Toit Papers—BC 722: C1—Research Papers Relating to his Book "Our Wandering Continents": C1.1 Review articles about his book "Our Wandering Continents": Malcolm Burr, "The Earth's Instability," in *Discovery*, March 1938.

[28] UCT-JL: Alex L. Du Toit Papers—BC 722: C1—Research Papers Relating to his Book "Our Wandering Continents": C1.1 Review articles about his book "Our Wandering Continents": "Continental Drift," in *The Geographical Review*, G.M. Wrigley (ed.).

[29] UCT-JL: Alex L. Du Toit Papers—BC 722: C1—Research Papers Relating to his Book "Our Wandering Continents": C1.1 Review articles about his book "Our Wandering Continents": "Continental Drift—Our Wandering Continents" (Author and Title Unknown).

in his other geological work. He also addresses the recalcitrance of scientists who, while disputing religious belief, are less inclined to be as flexible when their own preconceptions are challenged, attributing this to an intransigence that comes with age. Gevers also provided a criticism of the "contracting" and cooling Earth perspective that had been used to account for the geological features of the Earth's surface which had been contradicted by discovering radioactive decay of rocks which contributed to the heat of the Earth's interior. He acknowledged that Du Toit's use of this same radioactive heat as a possible mechanism of continental drift had the potential to address the great weakness of the theory—the lack of effective mechanism for continental movement. However, Gevers' use of the passive voice as he concludes served to effectively distance him from the theory: "By suggesting forces of sufficient magnitude and therefore a feasible cause of continental drift, it can be said that Dr du Toit ... has rendered considerable service to the theory he so ardently advocates".[30]

Despite Gevers' ambivalent response, continental drift had both its supporters and detractors and took on greater significance as well. For Du Toit South Africa was ideally placed—both geographically and ideologically—to take advantage of the opportunity created by continental drift theory to assert itself intellectually in a field that could have far-reaching implications for geology.[31]

Supporters

One of the creators of the theory, Frank Bursley Taylor, wrote a detailed letter after reading *Our Wandering Continents* in early 1938—a copy of which had been sent to him by Du Toit's publisher. Taylor was particularly appreciative of Du Toit's generous acknowledgement of Taylor's contribution to the theory: "Besides the beauty of the book and its binding, I am very much pleased at what I find inside. You certainly give me a full measure of recognition as the first man to put the idea of continental drift

[30] UCT-JL: Alex L. Du Toit Papers—BC 722: C1—Research Papers Relating to his Book "Our Wandering Continents": C1.1 Review articles about his book "Our Wandering Continents": T. W. Gevers "Earth's Continents are Drifting: Brilliant Exposition by a South African" [Newspaper unknown], Wednesday, 29 December 1937.

[31] T.W. Gevers. "The Life and Work of Dr. Alex. L. Du Toit"—Alex L. Du Toit Memorial Lectures No. 1. In *The Geological Society of South Africa: Annexure to Volume LII* (Johannesburg: The Geological Society of South Africa, The South African Association for the Advancement of Science, The South African Geological Society, 1949), p. 37.

on a sound scientific basis".[32] Taylor went on to defend his own work which had been focused on the more recent geological past as well as the incorporation of astronomy into his theory by crediting the action of the moon's gravitation pull for continental movement. It is an area that he wished to pay particular attention to should his health permit. Despite having not yet completed his reading of *Our Wandering Continents* due to having difficulty with his vision, he writes with enthusiastic approval of Du Toit's opus describing it as "as the record of a fine piece of scientific work in this relatively new field [that] must have involved a large amount of study and research" and promises to write Du Toit again as soon as he completed reading the book.[33] Taylor's words take on an added poignancy just six months later as Du Toit wrote a note of condolence to Taylor's widow in response to her sending him clippings of Taylor's obituaries that appeared in the American press. Du Toit was saddened—the deaths of first Wegener and Taylor meant Du Toit was no longer able to engage with the other pioneers in the field, leaving him to face the detractors on his own over a theory that evoked hostility. Their work had nevertheless laid the foundation for his own, just as his work had helped shaped their thinking: "In any further contributions I shall ever keep in mind the courageous actions of Taylor and Wegener to fathom the mystery of Earth Evolution".[34]

Yet Du Toit was not isolated.

Phillip H. Kuenen was more circumspect in his praise. Born in Scotland in 1902 but raised in Holland where he received his PhD in Geology at Leyden University, Kuenen came to specialise in marine geology and, in particular, turbidity currents which transported and deposited sediment on the ocean floor.[35] Kuenen's interest in marine geology was a result of his participation in the Snellius Expedition. The expedition, funded by the Dutch government both in Holland and the colonial government in the East Indies, took place between 1929 and 1930. Its aim was to investigate the sea around the East Indies and it contained specialists in

[32] UCT-JL: Alex L. Du Toit Papers—BC 722. C1—Research Papers Relating to his Book "Our Wandering Continents": C1.1 Correspondence (1935–1938) Letter by FB Taylor to Du Toit, 24 January 1938. UCT-JL.

[33] Letter by FB Taylor to Du Toit, 24 January 1938.

[34] UCT-JL: Alex L. Du Toit Papers—BC 722. B1—General Correspondence (1930s–1940s): Letter by Du Toit to FB Taylor's widow, 21 July 1938. UCT-JL.

[35] Francis P. Shepard, "Memorial to Phillip H. Kuenen, 1902–1976" (The Geological Society of America, 1978). ftp://rock.geosociety.org/pub/Memorials/v08/Kuenen-PH.pdf, p. 1. Accessed 15 September 2017.

oceanography, meteorology, botany and geography (Kuenen).[36] The expedition introduced Kuenen to the varied and dramatic geological landscape that characterised this part of the world and he subsequently published papers on diverse topics ranging from coral reefs to volcanoes.[37] Like his Dutch colleague, Waterschoot van der Gracht, it was this exposure to geology very different from that of largely stable Europe that allowed for a certain flexibility in entertaining the possibility of drift. Kuenen—who had been sent a copy of *Our Wandering Continents*—found no fault with Du Toit's argument but highlighted the old bugbear—the absence of an adequate mechanism to account for drift, without which continental drift lacked substance:

> I have just received your fine gift of "Our Wandering Continents" ... Your marsheling [*sic*] of the points in favour of drift make it even more probable to me that it has actually occurred ... My chief objections to the theory are that with the present geophysical state of affairs I see no possibility of drift from a mechanical point of view even if we had sufficient force. Your book has not done away with these difficulties. This is not an argument against the conception of drift, but against the proposed mechanism. I have the impression that we must still seek for a plausible mechanical picture, although it may be very long before we find it: there is still such a lot to be learnt about geophysics![38]

Detractors

Du Toit had, of course, anticipated the criticism of *Our Wandering Continents*—it was a variation of the theme that had dogged continental drift for decades. In his earlier thoughts on drift theory, he had suggested that this would be a challenge for conventional thinking in geology, a science that had traditionally progressed through controversy. Du Toit saw his work as the antithesis of the "orthodox": "For explanation I am advancing in all seriousness the view, revolutionary and heretical as it

[36] E. van Everdingen, "The Snellius Expedition," in *ICES Journal of Marine Science*, 1930, p. 320. https://academic.oup.com/icesjms/article-pdf/5/3/320/1825039/5-3-320.pdf. Accessed 15 September 2017.

[37] Shepard, "Memorial to Phillip H. Kuenen," p. 1.

[38] UCT-JL: Alex L. Du Toit Papers—BC 722: C1—Research Papers Relating to his Book "Our Wandering Continents": C1.1 Correspondence (1935–1938)—Letter by Ph. H. Kuenen (Geologisch Instituut, Rijks-Universiteit, Te Groningen) to Du Toit, 7 March 1938.

will appear to orthodox geologists". In 1924, speaking before the South African Association for the Advancement of Science, he contextualised continental drift within the history of geology, describing the discipline as "dogmatic", an example of which is the recalcitrance displayed over the age of the Earth, a controversy that persisted until the early twentieth century. For Gevers, Du Toit's position as a South African, a member of a young country and former colony, predisposed him to challenge accepted belief: "Here speaks a South African, to whom, in common with many representatives of other 'new' countries, authority, precedent, and convention mean little". For Gevers, too, it was the challenge presented by upstarts against the bastion of established thought and, with it, a certain respectability. Even contemplating the possibility of continental drift "led to reflections on the adequacy of one's decent bourgeois upbringing". For Du Toit, however, the matter was simpler—it was an hypothesis, an explanation for observations that could be subjected to rigorous scientific testing in order to determine its suitability.[39]

Du Toit had no small amount of trepidation regarding its reception as he worked on *Our Wandering Continents*, the work that would come to embody his thinking on continental drift. There were a number of potential sources of dissent and the controversy appeared to highlight a generation gap: "There seems to be a most determined, almost a vicious, unwillingness to permit the rift and drift heresy to grow, sort of 'They shall not pass!' attitude ... I would not have believed that supposedly truth-seeking scientific men could be so bitterly personal in a purely academic discussion. There are those in this country who have so lost their poise that they not hesitate to use their official positions to smother discussion".[40] As American geologist, Howard B. Baker pointed out, challenging the established views of the old guard could have adverse repercussions for the careers of younger geologists, making many unwilling to risk voicing their support for drift.

Already an established figure in his field, Du Toit was ready to tackle his potential detractors. He was aware of the need to make the work as comprehensive as possible in order to silence the critics and believed that his work would make better use of the geological stratigraphic evidence in

[39] Gevers, "The Life and Work of Dr. Alex. L. Du Toit," pp. 82–86.

[40] UCT-JL: Alex L. Du Toit Papers—BC 722: C1—Research Papers Relating to his Book "Our Wandering Continents": C1.1 Correspondence (1935–1938): Letter by Howard B. Baker to Du Toit, 10 August 1934.

support of drift than had hitherto been the case. He found a sympathetic audience in Baker as he acknowledged the difficulty of challenging the very basis of the field and the dogmatic views with which he had himself been inculcated in his early years. For Du Toit, *Our Wandering Continents* would be a throwing down of the gauntlet: "Whether the arguments will prove convincing I do not know, but I am at least making a valiant effort to undermine the fabric of current geology & so induce a feeling of uncertainty into orthodox minds. There is nothing like carry war into the enemies camp!"[41] And it was this adversarial attitude that would come to contextualise the book.

From an early simple advocacy of the scientific method and an entreaty to the "orthodox" to judge drift theory on its merits—and the substantial supporting evidence that he had amassed—Du Toit adopted a less placating tone in *Our Wandering Continents*. He once again drew upon the history of geology to demonstrate what seemed to be an innate resistance to new modes of thinking, describing American geology's view of stable and enduring continents and oceans as "archaic". He is particularly scathing of "land bridges": "For the words 'sink such land-bridges' we might with at least equal fairness substitute 'move the continents'". Yet the theory of land bridges was not built on the sturdy foundation of drift theory—the latter had been built on evidence—albeit circumstantial—which Du Toit had played a large part in assembling. The former theory, however, contained within it a convenient arbitrariness that did not imply a sense of "order", of structure. For Du Toit, then, it was only rational and scientific to abandon a static view of the Earth for a dynamic one[42]—but he faced formidable opposition, as unyielding as granite.

Some of those who were persuaded by *Our Wandering Continents*, such as Reginald Daly or Frank Bursley Taylor had already espoused the theory. It was more difficult to convince the others. In his biography of Du Toit, T.W. Gevers demonstrated a certain ambivalence towards drift theory, describing it as the most "elegant" explanation but only one among many. While emphasising Du Toit's great strengths as a geologist—his unwavering enthusiasm, exhaustive fieldwork and conscientious observations and recordings—Gevers appears to imply that Du Toit has

[41] UCT-JL: Alex L. Du Toit Papers—BC 722: C1—Research Papers Relating to his Book "Our Wandering Continents": C1.1 Correspondence (1935–1938): Letter by Du Toit to Howard B. Baker, 6 April 1936.

[42] Du Toit, *Our Wandering Continents*, pp. 4–7.

been premature in his early conclusions regarding drift theory. This is soon rendered unambiguous: "To my mind the host of features listed by Du Toit in this connection are highly interesting and suggestive, but do not in themselves constitute proof of much closer proximity in previous time". Du Toit's use of isostasy as a mechanism for drift is sharply dismissed as being "plausible [but] nevertheless completely in the realm of speculation".[43] With Du Toit's hypothesis summarily dealt with, continental drift was instead viewed as one hypothesis among many designed to explain the geological processes of the Earth. That Du Toit privileged drift theory was due to his individual "mental characteristics" that compelled him:

> to search for a unifying scheme and one grand controlling principle. The stronger this urge, the more readily will a person so endowed embark on a synthesising task, even when the foundation for such a superstructure is as yet incomplete. Du Toit very definitely preferred an *orderly* house to a shambles.[44]

Strong words from a colleague. By the end of his biography, Gevers had all but dismissed continental drift as a viable theory believing that, without Du Toit as its champion, it had "probably passed its zenith". Du Toit's strength lay not in promoting a soon-to-be defunct theory but in stimulating debate and demonstrating a breadth of original thought substantiated by extensive fieldwork.[45] Yet Gevers' ambivalence was relatively mild in relation to the storm of criticism that Du Toit's work provoked among geologists in Europe and, in particular, North America.

A particularly vociferous critic was palaeontologist George Gaylord Simpson. His criticism was confined to the fossil record and appeared in the *American Journal of Science* in 1943, drawing upon Du Toit's own espousal of a land bridge. Based on an analysis of fossilised teeth derived from an early species of horse, *Hipparion*, Leonce Joleaud, a French geologist, postulated a land bridge connecting Europe and Florida. Despite the new perspective offered by continental drift—which Joleaud came to believe—this land bridge was affirmed in 1929 by the president of the Geological Society of London. Alex Du Toit also considered the evidence

[43] Gevers, "The Life and Work of Dr. Alex. L. Du Toit," pp. 95, 41, 90, 97.

[44] Gevers, "The Life and Work of Dr. Alex. L. Du Toit," pp. 98, 101.

[45] Gevers, "The Life and Work of Dr. Alex. L. Du Toit," p. 102.

for this land bridge to be indisputable. If this land bridge, why not others? That formed the crux of Simpson's criticism—Du Toit's fossil evidence for continental drift was not irrefutable and he had had to occasionally resort to land bridges. To then postulate the dramatic movement of entire continents seemed unnecessary. Simpson suggested instead that the fossil record used to substantiate drift could be interpreted as the movement of animals north in reaction to a prospective ice age; their expedient arrival in Madagascar as a result of swimming or being carried along by floating detritus. Du Toit's response a year later was that Simpson's argument was flawed by his reliance on fossilised evidence derived solely from *Mammalia*, relative newcomers to the fossil record. A different picture would emerge if he extended his focus to include other, and more ancient, classes.[46]

In his definitive history of continental drift, Henry Frankel points out that *Our Wandering Continents* was not an accessible piece of work and that its effects on those opposed to drift or the "fixists" were negligible. One of those unpersuaded by mobilism was R.T. Chamberlin.[47] It was, ironically, the University of Chicago that would be at the forefront of the "New Geology" in the wake of the Second World War. The Department of Geology at the University of Chicago was established by Chamberlin's father, Thomas Chrowder Chamberlin and the younger Chamberlin would spend the greater part of his student and professional life associated with the institution. The establishment of the Geology Department came during a period of great change in American geology with the creation of the Geological Society of America and the U.S. Geological Survey as well as developments in mining, a better understanding of the geological history of the continent in relation to glaciation and the growing prominence of dinosaur fossils. This dynamism was reflected at the University of Chicago under the older Chamberlin who played a significant role in challenging the dominance of the revered eastern universities and instituted the *Journal of Geology*. This period of ebullient growth gave way to one of accumulating data within existing frameworks where "old ideas were warmed over and lost their freshness"—at least until the upheaval wrought by continental drift. This period of inertia coincided with R.T. Chamberlin's career as a geologist and, in many ways, this was a career that both paralleled and diverged from that of Du Toit. Chamberlin was particularly interested in chemistry and made significant strides in Du Toit's own area

[46] Wood, *The Dark Side of the Earth*, pp. 108–109.
[47] Frankel, *The Continental Drift Controversy I*, p. 306.

of expertise, glaciology, with much of his own research built on the earlier work of his father such as that of the movement of glaciers and on folding in the Rocky and Appalachian mountain ranges. While Chamberlin also carried out fieldwork in Florida, Brazil and the Pacific islands, he had little interest in geological mapping and preferred reaching conclusions based on existing data, experimentation and the use of analogy—the method where conclusions drawn in one area may be applied to another provided similar conditions are met. While undertaking research on the iron ore of Brazil, Chamberlin visited other areas in South America a decade before Du Toit. His interest in glaciation meant research trips to countries in the northern hemisphere and he was present at the International Geological Congress held in Pretoria in 1929. His conclusions were in direct contrast to Du Toit and, like his peers, Chamberlin held to the notion of fixed and perpetual continents and oceans, giving short shrift to the notion of continental drift.[48]

In his published response to *Our Wandering Continents*, Chamberlin wrote approvingly of Du Toit's amassing of evidence from a variety of disciplines, citing this as the great strength of the work with the qualifier, "irrespective of the conclusions expressed". At the same time, he considered Du Toit's reconstruction of continental motion to be arbitrary and designed to suit the circumstance rather than being based on scientific principle. He dismissed Du Toit's use of the convergence of stratigraphic, fossil and glacial evidence as unconvincing and open to alternative explanations—which he did not provide. His conclusion is an acknowledgement of Du Toit's "enthusiasm" for continental drift but it is, nonetheless, an enthusiasm which left Chamberlin unmoved.[49]

Chamberlin's reservations were echoed by Chester Longwell of Yale University. Born in Missouri in 1887, Longwell came to geology relatively late after spending his first seven years after completing high school working at a variety of odd jobs before completing a bachelor's and then master's degree from the University of Missouri. After a stint in the army during the First World War, he resumed his PhD at Yale University. His doctoral research in 1919 entailed fieldwork in conditions that were markedly similar to those endured by Du Toit. Longwell spent five months in

[48] F.J. Pettijohn "Rollin Thomas Chamberlin" (Washington, DC: National Academy of Sciences, 1970). http://www.nasonline.org/publications/biographical-memoirs/memoir-pdfs/chamberlin-rollin-t.pdf. Accessed 14 March 2017, pp. 89–91, 96, 99–100.

[49] Frankel, *The Continental Drift Controversy I*, pp. 306–307.

the Muddy Mountains of Nevada, an area so remote that it had barely been mapped. Lowell's main mode of transportation was by mule or on horseback and he found himself camping alongside prospectors or Native Americans.[50]

His research into geosynclines and the thrust faults of this region would mark a lifelong interest in geology and a continuation of fieldwork in this area until he was well past retirement. By 1920, Longwell had joined the Department of Geology at Yale University, eventually becoming its chair. His teaching led to an interest in orogeny or the means by which mountains are built. This would, of course, have introduced him to continental drift and he was to maintain an opposition to drift until as late as the mid-1950s.[51]

It was Du Toit's very enthusiasm that Longwell found particularly off-putting with his criticism at times taking on a more personal tone, "Perhaps the chief criticism of Du Toit's presentation is that it betrays somewhat too clearly the viewpoint of the zealous advocate".[52] For Longwell, geologists had progressed to the stage where they could be trusted to be impartial in their understanding of the merits of the drift debate. Du Toit disagreed, citing the implacable hostility of fixists (with Chamberlin singled out in particular) as provoking his "zealous" response, a tone that had raised the ire of those who retained their belief in the neutrality of science: "I admit the zealous presentation, but what can one do when confronted by minds of the R.T. Chamberlin type, as shown up in his review in the Journal of Geology ... Still, if geologists are really more open minded today, a more friendly & impartial tone will be assumed by me in future writings".[53]

Initially, Longwell proved himself willing to be persuaded, symbolic of the "open mindedness" urged by Du Toit. He acknowledged that he may have been premature in believing that geologists would simply be persuaded by a weight of evidence, leading them to discard the presumptions that they brought to continental drift. Longwell saw himself as representative of the impartial scientist and one who was, moreover, not unopposed to drift unlike his other American colleagues:

[50] John Rodgers, "Chester Ray Longwell" (Washington, DC: National Academy of Sciences, 1970). http://www.nasonline.org/publications/biographical-memoirs/memoir-pdfs/longwell-chester.pdf. Accessed 15 March 2017, pp. 250–251.

[51] Rodgers, "Chester Ray Longwell," pp. 250–252.

[52] Frankel, *The Continental Drift Controversy I*, p. 308.

[53] UCT-JL: Alex L. Du Toit Papers—BC 722: B3—Correspondence—B3.2 Correspondence, G-H: Letter by Du Toit to Chester Longwell, 2 January 1939.

> It may be that I am reflecting my own attitude, which certainly is altogether friendly toward the hypothesis. It seems to me that any fair-minded person who takes time to think the matter through will agree that the idea of horizontal continental displacements offers more promise of solving many of our large problems than any other suggestion that has been made.[54]

Initially—and apparently—in agreement with Du Toit that continental drift offered a compelling explanation for observed geological processes, Longwell's views were destined to shift as much as those of Du Toit's continents. By 1944, he criticised Du Toit's work for what he saw as discrepancies in the former's reconstruction of continental movement, considering the explanations to be "vague" and hardly "fundamental" as Du Toit had claimed.[55]

Like others before him, and dating back to the earlier work of Wegener, Longwell seized on the glaring weakness of continental drift theory—the untestability of the mechanics of motion which, in Du Toit's hypothesis, related to a combination of the Earth's rotation and the convection currents arising from the planet's interior as a result of the radioactive decay of elements within rocks and their subsequent release of heat—the latter being a prophetic but clearly premature view of the mechanisms of continental movement.[56] Nor was Longwell content to confine his dissent to private correspondence.

A further criticism aimed at Du Toit (and Wegener) by Longwell was over the measurement of drift itself. In a brief article appearing in the *American Journal of Science* Longwell disputed the motion in Greenland as postulated by Wegener and Du Toit of approximately 36 metres annually in a westerly direction. Subsequent measurements taken between 1927 and 1936 indicated that no movement was apparent and Longwell concluded, "At present, therefore, there seems to be no basis for any claim that longitude determinations in western Greenland support the concept of continental drift".[57] Dr Fritz Philipp Loewe of Melbourne University

[54] UCT-JL: Alex L. Du Toit Papers—BC 722: B3—Correspondence—B3.2 Correspondence, G-H: Letter by Chester Longwell to Du Toit, 13 February 1939.

[55] UCT-JL: Alex L. Du Toit Papers—BC 722: B3—Correspondence—B3.2 Correspondence, G-H: Letter by Chester Longwell to Du Toit, 1 July 1944.

[56] Frankel, *The Continental Drift Controversy I*, p. 308.

[57] UCT-JL: Alex L. Du Toit Papers—BC 722: C1—Research Papers Relating to his Book "Our Wandering Continents": C1.4—Geodetic: Chester R. Longwell "The Mobility of Greenland," in *American Journal of Science*, Vol. 242, Nov. 1944, p. 624.

explained this discrepancy to Du Toit by suggesting that the earlier calculations that supported drift were made using outdated observational methods whereas the new use of "wireless signals" in calculating longitude provided a far more accurate measurement, thereby discounting movement.[58] Du Toit's response was that, as he had pointed out in *Our Wandering Continents*, the argument for continental drift did not depend on the measurement of changes in longitude: "it fortunately happens that the geological & palaeoclimatological evidence is abundant & decisive, indicating without question that the crust composing the continents must have drifted across the face of the Earth with some rotation, & thus providing the reality of Continental Drift". The other evidence was sufficiently "weighty" and, while it could be disproved, Du Toit disarmingly put it, "that is for other persons to decide".[59] The fact that Longwell's criticism appeared years later must have seemed to Du Toit a means of rehashing old arguments that he believed he had sufficiently addressed. In the face of Longwell's hostility—and that of others—Du Toit remained adamant: "There is scarcely a geological journal picked up at random that I find does not contain something to support Drift, or weightier still, which does not oppose it, so why should I any longer doubt!"[60]

Another form of criticism—albeit of a more constructive type—related to Du Toit's writing style. Like his other published work, *Our Wandering Continents* was based on the extensive and detailed notes that Du Toit tended to accumulate while in the field. The emphasis is on description which is a characteristic of the final published version. In addition, Du Toit's desire to swing opinion in favour of drift meant that a wealth of detail was amassed which had the effect of overwhelming—if not bewildering—the reader. Du Toit's case was thus further harmed by the unwieldy nature of the work and its "tedious detail".[61] Unsurprisingly, *Our Wandering Continents* unequivocally excluded a lay readership. This lack of accessibility was diplomatically pointed out to Du Toit by Thomas

[58] UCT-JL: Alex L. Du Toit Papers—BC 722: C1—Research Papers Relating to his Book "Our Wandering Continents": C1.4—Geodetic: Letter by F. Loewe to Du Toit, 25 October 1937.

[59] UCT-JL: Alex L. Du Toit Papers—BC 722: C1—Research Papers Relating to his Book "Our Wandering Continents": C1.4—Geodetic: Letter by Du Toit to F. Loewe, 20 November 1937.

[60] UCT-JL: Alex L. Du Toit Papers—BC 722: B3—Correspondence—B3.2 Correspondence, G-H: Letter by Du Toit to Chester Longwell, 2 September 1944.

[61] Frankel, *The Continental Drift Controversy I*, p. 308.

Holland of Edinburgh University.[62] For his part, Du Toit acknowledged the criticism as well as his lack of talent for "popular writing", demonstrating a frustration with having to cater to two extremes of opinion—the one requiring an even more comprehensive explanation with an elaboration of the existing material and the other a cursory introduction designed to introduce the basic elements of drift theory.[63] It was a theme he reiterated in correspondence with Longwell, demonstrating the way in which Du Toit's decision to write *Our Wandering Continents* had been shaped in part by the hostility aimed at Alfred Wegener's earlier work: "Some of my friends have wanted a more popular treatment like that by Wegener, but then he was assailed on the score of incorrect facts & I have had to assemble a good deal of admittedly tedious data to give something for the critics to dispose of".[64]

The mixed reception of *Our Wandering Continents* is remarkable for the way it did little to alter the battle lines formed by the mobilists and the fixists. Du Toit's remarkable compilation of evidence that was breathtaking in its scope and ambition—which bore some responsibility for the density of his prose—while, being enthusiastically received by the mobilists, failed to convince his detractors. It is perhaps an indication of the lack of impartiality among many geologists that, prior to the explosion of evidence for plate tectonics, only a few American geologists were persuaded by *Our Wandering Continents* to alter their fixist views.[65] The reception of the theory thus exemplified the contextualisation of scientific thought. As Bailey Willis of Stanford University put it: "I recognize, of course, that each of us is the victim of circumstances, and that our convictions are more or less imposed upon us by bend of mind and experience. I therefore shall not dispute with you the question of whether or not continents can drift but would assure you that I appreciate the courage with which you have set forth the state of the problem. You certainly have succeeded in putting the best possible face upon your solution of it".[66]

[62] UCT-JL: Alex L. Du Toit Papers—BC 722: B3—Correspondence—B3.2 Correspondence, G-H: Letter by Thomas Holland to Du Toit, 9 November 1938.

[63] UCT-JL: Alex L. Du Toit Papers—BC 722: B3—Correspondence—B3.2 Correspondence, G-H: Letter by Du Toit to Thomas Holland, 2 December 1938.

[64] Letter by Du Toit to Chester Longwell, 2 January 1939.

[65] Frankel, *The Continental Drift Controversy I*, p. 309.

[66] UCT-JL: Alex L. Du Toit Papers—BC 722: B1—General Correspondence (1930s–1940s)—Letter by Bailey Willis to Du Toit, 24 November 1937.

Characterised by a distinctive white and bushy beard, Bailey Willis—like the other geologists of his generation—experienced the vagaries of geological fieldwork in the uncertain terrain of the nineteenth century with a special focus on the Appalachian mountain range. Like Du Toit, he had also received a grant from the Carnegie Institution that allowed him to investigate the geology of northern China in 1903—at that point, largely unknown to geologists. His interpretation of his findings was made within the confines of prevailing geological doctrine shaped by uniformitarianism that included erosion and the "warping" of the Earth's crust that shaped landforms. Described as being "often ready with an explanation long before he had the facts necessary to establish his opinion", Willis was nevertheless ardently opposed to continental drift theory, forming—along with Chamberlin and Longwell—part of the vanguard of the fixists.[67]

Two years after its publication, Du Toit once again found himself having to defend continental drift, this time in a debate with Arthur Coleman, a prominent Canadian geologist who had obtained his PhD from the University of Breslau in Germany and was Professor Emeritus at the University of Toronto.[68] Describing himself as an "old-fashioned geologist", Coleman heaped criticism on Du Toit's continued defence of mobilism. After the mixed reception received by *Our Wandering Continents*, Du Toit published work on the genesis of the Arctic and Atlantic Oceans with an emphasis on their significance for the southern hemisphere continents. He wrote a similar piece on the Pacific Ocean as well and much of his later work focused on responding to the continuing criticism. Coleman's criticism centred on glaciation and, again, the lack of an adequate mechanism explaining drift. With regards to the former, he was sceptical that the arid continent that Gondwana was likely to be could not be a source of glaciation as claimed by Du Toit. In terms of explaining continental drift, Coleman refused to accept the parallels of floating continents with that of icebergs adrift on the ocean, "Permanently enclosed rock-masses cannot drift—they must be pulled or pushed".[69] In addition, from his perspective, there seemed to be no plausible explanation for the

[67] Eliot Blackwelder, "Bailey Willis: 1857–1949" (Washington, DC: National Academy of Sciences, 1961). www.nasonline.org/publications/biographical-memoirs/memoir.../willis-bailey.pdf. Accessed 11 September 2017, pp. 334–336, 341, 340.

[68] "Arthur P. Coleman—Biographical Sketch," Victoria University Library Special Collections, University of Toronto. http://library.vicu.utoronto.ca/collections/special_collections/f7_arthur_p_coleman. Accessed 15 March 2017.

[69] Frankel, *The Continental Drift Controversy I*, p. 310.

fragmentation of Gondwana to begin with. And, finally, echoing the sentiments of Chamberlin:

> The Gondwana plants are mainly ferns, equisetums and club-mosses, especially ferns—all cryptogams the spores of which could easily be transported by the wind. A gale would quickly carry them hundreds of miles. Why send continents crashing through the solid earth's crust to effect their distribution?[70]

Clearly, continents would move before the opposition was likely to. In response, Du Toit focused on glaciation, pointing out the possibility of water inlets which would bring the required moisture and also demonstrating that glaciation in the northern hemisphere had occurred at a greater distance from the coast than that attributed to Gondwana. He defied Coleman to find an alternative "fixist" explanation for the clear evidence of glaciation occurring simultaneously across distinct continents at different latitudes. A war-time response to Longwell, while acknowledging his own less than disinterested position, described the former as a fence-sitter who needed to consider that "fixist" and "mobilist" were either-or positions. Since all evidence pointed to the fact that continents were not, in fact, "fixed" the only alternative was that they moved. It was a logical deduction that did not require a "wait-and-see" approach.[71]

Of great interest to Du Toit was work that the geologists tended to ignore—Wegener's consideration of palaeoclimates and, in particular, that related to glaciation. It was indisputable that geologists [Du Toit himself one of them] had found tillite in Africa, the lithified remnants of glacial activity, "'Either geologists are suffering from delusions when interpreting those formations or else the Pole of those times wandered somewhere in or near that continent'".[72] And, if not wandering poles, then wandering continents were the only solution. This would reach some kind of resolution a decade later with palaeomagnetic research.

As Du Toit reached the end of his life, he was still unwilling to let *Our Wandering Continents* go, focusing on new ways of responding to his detractors. His attention turned to mathematics as a solution, a means of demonstrating statistically the likelihood of drift. An unpublished work

[70] Frankel, *The Continental Drift Controversy I*, pp. 310–311.

[71] Frankel, *The Continental Drift Controversy I*, pp. 311–312.

[72] Frankel, *The Continental Drift Controversy I*, p. 314.

dated from 1944–1945 "On the mathematical probability of drift" is evidence of this as is his own copy of *Our Wandering Continents* with extensive notes suggesting further revision that would use mathematics to substantiate theory.[73]

In Du Toit's biographical tribute to Arthur William Rogers, it is possible to see how Du Toit views both himself and the discipline of geology. He pays tribute to Rogers for his dedication to geology as a scientist—key to Rogers' work was the furthering of scientific knowledge and other considerations such as those linked to mining and economic geology, for instance, were secondary. At the same time, Du Toit draws a distinction between himself and the more "orthodox" Rogers who "tended to curb the exuberant and more commonly heterodox ideas of" Du Toit. The latter was then the rebellious sceptic while Rogers was a product of the establishment, "As might has [*sic*] befallen anyone educated in a centre where respect is even yet paid to tradition, and doctrines of geology are apparently not less sacred than those of theology".[74] Du Toit's words are thus a defence of his own work and willingness to challenge convention when measured against the more "traditional" Rogers—and the other South African geologists who were indifferent to his work on drift. His tribute to Rogers functions as harsh criticism of those who opposed drift as the origins of modern geology lay in the challenge it presented to biblical understandings of the creation and age of the Earth. Geology was a product of the Enlightenment, of the bringing to bear of scientific reason and methodology to understanding the mechanisms of the natural world yet Du Toit suggested that what was almost heretical in its youth could become an unyielding orthodoxy in its middle age, suspicious of change and challenge.

Alex Du Toit's death in 1948 left the question unresolved and the debate continued into the 1950s with new actors entering the scene while older detractors sounded both weary and patronising towards a theory that would not die. As Roger Wood put it, "this was trench warfare long after any enthusiasm for battle had expired". British geophysicist Harold Jeffreys persisted, citing elementary knowledge of the solidity of the mantle that did not allow for movement, concluding with the need for simple

[73] Frankel, *The Continental Drift Controversy I*, p. 314.

[74] Alex L. Du Toit, "Arthur William Rogers," in *South African Journal of Geology*, Vol. 49, pp. 291–304, 1946. http://www.sabinet.co.za. AJA10120750_1936. Accessed 13 October 2016, p. 297.

observation and description without the drive to fit these into a theoretical model. The battle lines were drawn—geophysics and the "orthodox" geology on one side and the fossil record and stratigraphy of the continents of the southern hemisphere on the other. Once again, at the heart of the position adopted lay space. Those who opposed continental drift came from seismologically stable regions of the northern hemisphere where their training emphasised stability and uniformity—Britain, France and the eastern United States. Yet new allies for drift came from an even more geologically quiescent part of the world—the Netherlands. It was the very inactive nature of their geology that sent them beyond the national borders that confined the British, Germans, French and Americans. Confronted by the Alps, they empathised with Suess and their colonies offered an even strong illustration of drift. In 1926 Dutch geologist, Waterschoot van der Gracht spoke of the support within Dutch geology for Wegener's hypothesis based on their observations in the Dutch East Indies, "'Without knowing why, we *see* that New Guinea drifts violently to the north'".[75] Another prominent Dutch geologist who had close links with South Africa as a "state geologist" in the Transvaal and who had worked with David Draper was Gustaaf Molengraaff. He had been part of geological expeditions to the East Indies and had become a prominent supporter of drift theory. As early as 1916, Molengraaff proposed that the Mid-Atlantic Ridge was created by the split of Europe and Africa from North and South America. He would later apply the same hypothesis to the East African Rift Valley. Molengraaff did, however, disagree with Van der Gracht's belief that the mechanism for drift was radioactivity.[76] Du Toit was more sympathetic to Van der Gracht's view and also paraphrased Van der Gracht in the Preface to *Our Wandering Continents* where he echoes the Dutchman's "plea for a little more tolerance of spirit from critics" when confronted with statements that had yet to be adequately proven.[77]

This is a theme that Van der Gracht elaborated upon in his correspondence with Du Toit and in the wake of holding a symposium on continental drift in 1928. The symposium was as a result of the controversy aroused

[75] Wood, *The Dark Side of the Earth*, pp. 110–111.

[76] Frankel, *The Continental Drift Controversy I*, pp. 197–198. The importance placed on radioactivity by Van der Gracht was based on the work of John Joly which is addressed in the Postscript.

[77] Du Toit, *Our Wandering Continents*, p. viii.

by Wegener's publication and was initiated by the American Association of Petroleum Geologists. The publication arising out of the symposium contained contributions by both the mobilists and fixists—among them figures that would both support and criticise Du Toit. Included was Wegener himself, Frank Bursley Taylor and Waterschoot van der Gracht. The vehement opposition was largely American and included Charles Schuchert, Bailey Willis and Chester Longwell among others. The conclusion of the symposium highlighted the predicament of scientists faced with a tectonic shift in thinking: "This entire symposium stands as a token of the insufficiency of what we have been teaching in the past to explain the facts, and also, of our inability to make our new attempts at an explanation fit in with everything which we now believe to be a fact either in geophysics, geology or biology".[78] While Van der Gracht was portrayed as a proponent of drift theory, he saw himself as being unmoved by the extremes of opinion that had come to characterise the debate, especially on the part of American geologists who he considered to have an almost knee-jerk reaction: "Their attitude was very much to waive the whole question aside as 'an impossibility' and something outrageous; they were so cocksure about it that they did not even take the trouble to study the matter properly, to know what it was really all about, and which questions were involved. Their teaching to their students was all in that spirit".[79]

It would be an oversimplification to state that all southern hemisphere geologists were ardent supporters of continental drift. Many Australian geologists, in particular, opposed mobilism. An exception was Samuel Warren Carey. Born in New South Wales in 1911, Carey's interest in science was bolstered by the Great Depression of 1929. Initially, hoping to study medicine, science proved to be the more economic option and, at first year level, Carey took geology as an optional science course at the University of Sydney. It was during this period that Carey was first exposed to notions of continental drift. One of his geology professors, L.A. Cotton had written an article that considered polar wandering and also

[78] Review of "Theory of Continental Drift: A symposium on the origin and movement of land masses both inter-continental and intra-continental, as proposed by Alfred Wegener in Quarterly Journal of the Royal Meteorological Society," Vol. 55, Issue 229, January 1929, p. 102. Accessed from http://onlinelibrary.wiley.com/doi/10.1002/qj.49705522923/pdf, 11 September 2017.

[79] UCT-JL: Alex L. Du Toit Papers—BC 722: B2—Correspondence with colleagues in South America (ca. 1920s–1931)—Letter by Waterschoot van der Gracht to Du Toit, 20 August 1928.

emphasised the importance of continental drift as an explanation for the geological anomalies of the southern hemisphere. A recently retired professor and significant influence on Carey's development as a geologist, Edgeworth David, also published a paper on continental drift in 1924.[80] David had had an adventurous career as a geologist—an expert on glaciation, he had also been part of Ernest Shackleton's expedition to Antarctica reaching the South Magnetic Pole. He had met Alex Du Toit in 1914 when the latter visited Australia and both men had observed the Paleozoic era glacial rock strata. In 1926, David returned the visit, meeting with Dr Toit in South Africa and was given a preview of *A Geological Comparison of South America with South Africa*. However, David had already retired and was nearing the end of his career. His lecture "Drifting Continents: The Wegener Hypothesis" was delivered in 1928 when he was 70.[81] He nevertheless made a tremendous impact on Carey. Continental drift thus became an interpretative lens through which Carey conducted his fieldwork in New Guinea and formed an integral part of much of his later academic work. Mirroring the eccentric interest of Du Toit, Carey's first published paper focused on water dowsing.[82]

Like Du Toit, Carey was intensely involved in geological mapping from his Honours and, to obtain his master's degree in geology, he carried out mapping of the Werris Creek region. His special focus was on Carboniferous and Permian stratigraphy—both key periods in the history of Gondwana. With a keen interest in fieldwork and an expertise in structural geology, Carey accepted employment at Oil Search Ltd. and was sent to work in New Guinea. New Guinea is geologically active with major tectonic activity which Carey was able to witness first-hand and which formed the basis for his DSc with his thesis, "Tectonic Evolution of New Guinea and Melanesia" completed in 1939.[83]

After active service in the Second World War, Carey took on the role of Government Geologist of Tasmania, eventually becoming Foundation Professor at the University of Tasmania's new Department of Geology in 1946 where he introduced his students to plate tectonics. To reconstruct continental movement, Carey had a model of the southern hemisphere

[80] Patrick G. Quilty and Maxwell R. Banks, "Samuel Warren Carey 1911–2002." Australian Academy of Science. https://www.science.org.au/fellowship/fellows/biographical-memoirs/samuel-warren-carey-1911-2002. Accessed 15 March 2017.

[81] Frankel, *The Continental Drift Controversy I*, pp. 497–498.

[82] Quilty and Banks, "Samuel Warren Carey 1911–2002."

[83] Quilty and Banks, "Samuel Warren Carey 1911–2002."

built and had detailed drawings of the continents which he then proceeded to move over the model based on existing palaeomagnetic, geographical and geological data. Through this process of experimentation he was able to refute Harold Jeffreys' claim that there was no real fit between the South American and African continents. In addition, Carey made contributions to understandings of rift valley formation as well as seafloor spreading. Another major contribution was the Continental Drift Symposium held in Hobart, Tasmania in 1956 yet just two years earlier Carey was not given a Fellowship at the newly formed Australian Academy of Science due to, what he believed, was his espousal of continental drift.[84]

Among those who Carey invited to the symposium in Hobart was Lester King of the University of Natal who, after Du Toit's death, was the leading South African proponent of drift. Also included was Chester Longwell who, in his opening address and closing remarks, was able to present a balanced view of the controversy with a particular consideration of Carey's own research into the connection between South America and Africa. The symposium also included other "fixists". R.A. Stirton, a graduate of the University of California, discarded the use of land bridges as a means of accounting for faunal similarity, instead suggesting that monotremes such as the forebears of the platypus had reached Australia from Southeast Asia as unwitting passengers on mangrove roots. A.H. Voisey of the University of New England, New South Wales believed that, rather like the formation of a pearl in an oyster, continents were a result of the addition of mountainous land. While unable to provide an explanation for this accretion, he nonetheless criticised the supposed fit between continents emphasised by drift theory, finding it arbitrary.[85]

The most vehement Australian critic of drift was E.C. Andrews who had also studied at the University of Sydney under Edgeworth David but prior to David's own conversion to drift. A respected geologist who had been part of the Geological Survey of New South Wales and was president of the Linnaean Society of New South Wales, Andrews believed Wegener's theory to be a fable with little scientific merit. He dismissed the use of *Glossopteris* as evidence, hypothesising that the origins of the plant had most likely been Antarctica and, as Antarctica cooled, *Glossopteris*

[84] Quilty and Banks, "Samuel Warren Carey 1911–2002."

[85] Henry R. Frankel, *The Continental Drift Controversy: Volume II: Palaeomagnetism and Confirmation of Drift* (Cambridge, Cambridge University Press, 2012), pp. 335–336, 338–339.

had moved further north to Australia, Africa, India and South America. He did not, however, account for the movement of *Glossopteris* to landmasses divided by ocean. For Andrews and other Australian "fixists", Australia was the result of the accretion of smaller territories—the continent had thus come into being as a result of growth rather than fragmentation. Prominent geologist E. Sherbon Hills of the University of Melbourne also highlighted Australia's unique fauna which had clearly evolved in isolation and seemed to demonstrate no biological links to the fauna of other continents. Hills would only concede the existence of plate tectonics in 1970.[86]

Southern Assertion

A possible explanation for the easier acceptance of continental drift theory in the southern hemisphere, exemplified by Du Toit in South Africa and Juan Keidel in South America, may have been their geographical proximity to the evidence for Gondwana. What was perceived to be the overwhelming stratigraphic and fossil evidence suggested no other conclusion for these southern geologists.[87] Perhaps, too, there was no need of the ideological leap that would assert the prominence of the marginalised southern landmasses.

Yet, as Chakrabarti observes, "Geology is an imagination of the past based on the analogy of the present".[88] He points out that the way in which Gondwana was envisaged by geologists was a product of the social and cultural context of British imperialism. The conceptualisation of the ancient landmass of Gondwana—named for an Indian tribe, the Gonds—was based on the convergence of both "primitive" peoples and a primordial landscape. Not just confined to India, this vision incorporated Australia and its aborigines and the Karoo and the San. As was evident in South Africa, the discipline of geology could not be divorced from anthropology and history.[89] As Gondwana became a product of the British colonial imagination so, too, in the twentieth century, did it become a metaphor for nationalism.

[86] Frankel, *The Continental Drift Controversy I*, pp. 503–504, 505–507.

[87] Philip Stone, "Geological exploration of South Atlantic islands," p. 17.

[88] Pratik Chakrabarti, "Gondwana and the Politics of the Deep Past," in Past and Present, 242:1, February 2019, p. 123.

[89] Chakrabarti, "Gondwana and the Politics of the Deep Past," pp. 120–121.

Just as the theory of continental drift was contextualised by struggles over space both within South Africa in terms of physical space and segregation and with the northern hemisphere and the transmission of knowledge, it also allowed for the building of bonds across nations. The reconstruction of Gondwana, in particular, and the evidence found across the southern continents—most of which shared a similar history in terms of colonisation and a nascent nationhood—allowed for intellectual co-operation.

In 1914, Du Toit had travelled to Australia and in 1923 it was his trip to South America that placed him at the forefront or firing line of the continental drift debate. And from December 1937 to January 1938—in the wake of *Our Wandering Continents*—there was a further trip to India. By the end of his career, he was to have visited all the landmasses that comprised Gondwana with the exception of the Antarctic. These visits complemented a vigorous correspondence between geologists marked by both agreement and dissent and co-operation across national, racial and cultural boundaries.

One of these was S. Warren Carey of the University of Tasmania who was himself criticised for his espousal of drift theory and whose view of drift was modified from Du Toit as he "[had] been approaching the problem from the Australian viewpoint". This led him to a slightly altered vision of Gondwana and is tantalisingly suggestive of the way in which science was shaped by space. In the same letter, Carey went on to request Bokkeveld fossils from Du Toit to be used as a means of comparison.[90]

Du Toit's travels in South America also led to the formation of intellectual, collegial and not so collegial ties that served to metaphorically replicate the joining of the continents. Kenneth Edward Caster was an American palaeontologist who received his doctorate from Cornell University in 1933. Despite his northern roots, Caster, like Du Toit, was to develop a strong interest in the geology of the southern hemisphere landmasses. He was awarded a Guggenheim Fellowship allowing him to serve as Professor of Geology for three years at the University of Sao Paulo from 1944—where he was engaged in correspondence with Du Toit.[91]

[90] UCT-JL: Alex L. Du Toit Papers—BC 722: B1—General Correspondence (1930s–1940s)—Letter by S. Warren Carey to Du Toit, 10 November 1947.

[91] University of Cincinnati "Memorial: Kenneth Edward Caster (1908–1992)." http://homepages.uc.edu/~huffwd/Department_History/Kenneth_E_Caster/Caster_Memorial.htm. Accessed 1 September 2017.

Proficient in both English and Portuguese, Caster was instrumental in having *A Geological Comparison of South America with South Africa* translated into Portuguese. Du Toit readily agreed to the project. As he pointed out in the Foreword to the new edition, it would allow South Americans to understanding the geology of South Africa: "To me indeed there was the unrivalled opportunity for spreading some knowledge of the geology of the sister-continent of Africa, and incidentally of the Hypothesis of Continental Drift, the collective evidence of which becomes stronger year by year".[92]

In Caster, too, Du Toit also found a voice of empathy. Initially not yet a convert of continental drift, Caster commiserated over the hostility with which the theory was received, summarising the personalities of the key American detractors: "While I am neither pro or con, I am very much open-minded; with, in my humble opinion, the Wegnerian arguments ever growing stronger, I must confess the kind of argument that Dr. [Bailey] Willis sometimes indulges in displeases me. Simpson is much more seriously to be considered, for he is always open to conviction in whatever argument. Longwell is also a good man, but tends to speak a bit professorially on occasions (some of us call it the 'Yale complex' in the States)".[93] Aided by a second Guggenheim Fellowship, Caster extended his period in South America, investigating the geology of Brazil and serving as Visiting Professor at the School of Mines in Colombia. He later extended his geological research to New Zealand and South Africa and ultimately Australia where he was a Fulbright Visiting Professor at the University of Tasmania which would put him in contact with Carey. The opportunity for extensive travel and geological study served to persuade Caster of continental drift which would underpin his teaching and writings in the 1950s. As a result, he was awarded the Gondwana Medal of the Geological Survey of India in 1956.[94]

The final region that was significant in the reconstruction of Gondwana was the Indian subcontinent and Du Toit had carried on a correspondence

[92] UCT-JL: Alex L. Du Toit Papers—BC 722: B1—General Correspondence (1930s–1940s): B3.1 Correspondence mainly with Dr. E. Caster, 1941, 1945–1947—Draft of Foreword by Du Toit, 10 April 1946.

[93] UCT-JL: Alex L. Du Toit Papers—BC 722: B1—General Correspondence (1930s–1940s): B3.1 Correspondence mainly with Dr. E. Caster, 1941, 1945–1947—Letter by Kenneth E. Castor, Departamento de Geologia e Paleontologia, Universidade de Sao Paulo to Du Toit, 26 March 1946.

[94] University of Cincinnati "Memorial: Kenneth Edward Caster (1908–1992)."

with Indian scientists even prior to his visit in late 1937. Politically, the Indian subcontinent was a contrast to the independent settler states of Australia and South Africa as well as those in South America. By 1937—and for the next decade—India would remain a British colony and one that had had a tense and long history with South Africa. The descendants of Indian indentured labourers and merchants who had formed part of a global economy in the mid-nineteenth century were a significant component of the South African population as was their agitation for political rights. The man who would become an Indian nationalist icon, MK Gandhi, had cut his teeth in the embryonic political struggles of South African Indians, developing the skills that would be used in India's own struggle for independence. What is remarkable about Du Toit's correspondence with Indian scientists, however, is a sense of equality and scientific co-operation across racial lines that was notably absent from South African society at the time.

One such figure was Birbal Sahni, a palaeobotanist who had, from an earlier age—and with the reinforcement of his father, a professor of chemistry developed an interest in natural history, exploring the Himalayas and collecting rock, fossil and plant specimens. Like Du Toit, Sahni was the product of an education shaped by the Enlightenment and the British Empire. He received his Bachelor of Science degree from the University of Punjab and subsequently studied at Cambridge and the University of London. It was whilst at the former that Sahni came under the influence of the renowned A.C. Seward, a significant contributor to the fields of both geology and botany. Under Seward's tutelage, Sahni went on to research and publish prolifically and was also introduced to the flora of Gondwana when collaborating with Seward on the "Revision of Indian Gondwana Plants" published in 1920. Despite holding the position of Professor of Botany at Lucknow University, Sahni retained a passion for fieldwork and in 1923 his discovery and subsequent analysis of the fossil of *Glossopteris angustifolia* suggested that this important Gondwana remnant was seed-bearing rather than a fern as had been the conventional thinking. Fossil discovery followed fossil discovery and Sahni was an acknowledged expert on the prehistoric floral history of the Indian subcontinent. And, as with Du Toit, the fossil evidence—particularly that of *Glossopteris*, a plant characteristic of a colder climate—suggested the different position of the southern landmasses. A subsequent discovery in China of *Gigantopteris* flora—characteristic of a tropical climate—suggested to Sahni that India had initially been part of Pangaea and, upon the break-up of the supercontinent, had come into contact with the main part of present-day Asia.[95] This prompted among others, the publication of "The relations of the Indian Gondwana flora with those of Siberia and China" (1935), "The *Glossopteris* flora in India" (1935) and work that engaged directly with Wegener's notion of drift, "Wegener's theory of continental drift in the light of palaeobotanical evidence" which was published in *Nature* in 1936 and "Speculations on the climates of the Lower Gondwanas of India" delivered at the International Geological Congress held in Moscow in 1937. Unsurprisingly, then, Du Toit viewed Sahni as a significant contributor, adding to the overall picture of drift by unearthing evidence from what Du Toit termed "the 'border land' regions", that is, Asia and the Indian subcontinent, that would ultimately refine Du Toit's own understanding.[96]

This pattern of co-operation and the accumulation and assimilation of data was one that Du Toit would follow until the end of his career with correspondence to Dr D.N. Wadia, author of the *Geology of India* regarding the actual positioning of Sri Lanka (Ceylon) in relation to Gondwana.[97] In many ways, Wadia's career paralleled that of Du Toit with him achieving a similar prominence in Indian geology. The Wadia family had been deeply influenced by the scientific and technological development associated with British imperialism. A relative, Ardeseer Cursetjee, was an engineer and architect who was fascinated by the possibilities of steam and in the 1830s obtained a steam engine from Britain which he subsequently used to power a 60-ton ship. A technophile, Cursetjee can also be credited for importing sewing machines, gas lighting and photography and subsequently became the first Indian to be elected a Fellow of the Royal Society in 1841. One of the most visible signs of British imperialism in India was the network of railways that crisscrossed the subcontinent and Wadia's father worked for the Bombay, Baroda and Central Indian Railway. The young Wadia's interest in science was fostered by an older brother and he

[95] H. Hamshaw Thomas. "Birbal Sahni, 1891–1949," in *Biographical Memoirs of Fellows of the Royal Society*, 1 November 1950, Vol. 7, Issue 19, p. 270. http://rsbm.royalsocietypublishing.org/content/7/19/264. Accessed 5 September 2017.

[96] Thomas. "Birbal Sahni, 1891–1949," pp. 265–266, 267–268, 270, 275.

[97] UCT-JL: Alex L. Du Toit Papers—BC 722: B1—General Correspondence (1930s–1940s). Letter by Du Toit to Dr. M.S. Krishnan, Geological Survey of India, Calcutta, 14 November 1946.

subsequently went on to Baroda College where he graduated with a master's in biology and geology and began lecturing undergraduates.[98]

As was evident in South Africa, geological exploration was an important aspect of the imperial project and had been first undertaken during the reign of the British East India Company with the formation of the Geological Survey of India under the leadership of Thomas Oldham. This marked the beginning of a lengthy investigation of the geology of the Indian subcontinent, with a collation of the results. This was accompanied by the collection of geological specimens and their subsequent display at the Indian Museum in Calcutta, along with the institution of geology as an academic discipline at tertiary level. As a geologist, Wadia's career spanned the decades from colonialism to independence. This was symbolised by his lengthy tenure at the Prince of Wales College in Kashmir. Initially named to honour King George V, it was subsequently renamed Mahatma Gandhi College. Early in his career, Wadia had also bemoaned the lack of adequate teaching material for geology students with an Indian focus, prompting him to write the *Geology of India for Students* which was published in 1919 and was still a standard reference work four decades later.[99] Like Du Toit, he had begun his career with extensive fieldwork, helping map the western regions of the Himalayas. A description of Wadia in the field eerily echoes that of Du Toit:

> For months on end he would leave his camp at day-break for a good 20 miles traverse on foot in trackless mountains or a much longer ride on mule or horse-back, a late lunch, his first meal of the day ... In 1942 he was reported to the police in Namunukala, Ceylon as "a lonely khaki-clad figure silently hitting every stone in the neighbourhood."[100]

Wadia's fieldwork was not confined solely to the mainland and he had served as Government Mineralogist for Sri Lanka with particular focus on the country's geological resources necessary to develop a fledgling economy. Also reflecting Du Toit's eclectic career path, Wadia had been involved in mapping under the Indian Geological Survey, economic geology with the Bureau of Mines and, within the context of the nuclear age

[98] C.J. Stubblefield "Darashaw Nosherwan Wadia, 1883–1969," in *Biographical Memoirs of Fellows of the Royal Society*, 1 November 1970, Vol. 16, p. 544. http://rsbm.royalsocietypublishing.org/content/7/19/264. Accessed 5 September 2017.

[99] Stubblefield, "Darashaw Nosherwan Wadia," pp. 544–546.

[100] Stubblefield, "Darashaw Nosherwan Wadia," p. 556.

and conflict with Pakistan, had served as Geological Advisor to the Indian Atomic Energy Commission. He had also straddled the border between his prioritising of the needs of the nation versus the importance of fostering international co-operation and, as pointed out in his biography, his "nationalist views were tempered by his internationalist outlook". With regards to continental drift, Wadia's focus was on Kashmir in northern India which, while not demonstrating the evidence for glaciation, had provided evidence that it had been surrounded by glacial ice during the Permian and Carboniferous periods. The evidence prompted Wadia to suggest that Kashmir had provided a point of connection between Gondwana and present-day Eurasia that had allowed for the movement of flora between the two landmasses. As Indian independence approached, Wadia highlighted the value of science and technology in making use of natural resources that would foster economic growth and development. Geology and exploration would play a significant role in this—especially as access to existing resources was curtailed due to Partition. It was thus the location and extraction of valuable minerals that were to form the basis of the remainder of Wadia's career.[101]

The varying levels of interaction between these southern geologists—their tensions and collaborations—demonstrate a different path to the transmission of knowledge than a unidirectional movement from north to south.[102] For these geologists, operating within the scientific hegemony of Europe and North America, continental drift became the means by which southern and, particularly, national identities were asserted even as cross-border collaboration allowed for the accretion of data to support continental drift. It was also drift theory itself that provided a vehicle for the articulation of these various and sometimes competing identities and assertions of space. The recreation of past continents provided an alternative view to the dominance of the northern hemisphere and the preponderance of evidence derived from the south presented an opportunity for an assertion of centrality on the part of landmasses that had since been subject to imperial domination and exploitation. Yet new discoveries were simultaneously articulated within the context of a scientific method produced by the European Enlightenment, highlighting the complex ways in which knowledge was both produced and interpreted. At the same time,

[101] Stubblefield, "Darashaw Nosherwan Wadia," pp. 543, 552.

[102] While the Indian subcontinent is located in the northern hemisphere, it forms part of the "Global South" and comprised part of the southern continent of Gondwana.

despite being contextualised by varying levels of inequality and by both dependence and independence, drift would ultimately contribute to a sense of "holism" and of unity.

Alex Du Toit was certainly far-sighted in seeing the potential of Wegener's work and spent a large part of his professional career assembling the evidence to support it. Both the history of the theories of continental drift and plate tectonics, however, provide valuable insight into the contextualisation of scientific thought. The varying degrees of opposition faced by Wegener demonstrate the sometimes torturous process by which new—and challenging—perceptions of the world are received. The support provided by Du Toit was insufficient in the face of hostile criticism as Du Toit, based on a southern continent, isolated from the centres of hegemonic western scientific and intellectual development, was, ironically, a victim of geography. In addition, Du Toit's espousal of continental drift was not simply a concession to the strength of Wegener's work but was situated within an ideological milieu reinforced by figures such as Smuts and Hofmeyr. Gondwanaland literally positioned Africa at the geographical centre, challenging the political, cultural and economic dominance of Europe. It converged with the assertive patriotism of a new nation and catered to its imperial ambitions in Africa.

The movement of the continents challenged the ground on which we stand. Gondwanaland showed the artificiality of boundaries and borders. This was reinforced by discoveries in palaeoanthropology showing hominin migrations out of Africa and around the world. This converged with Smuts's own theory of Holism. Yet segregation, premised on racial inequality, continued in South Africa, and would reach its zenith under the apartheid state. Continental drift thus showed the ways in which scientific "truths" could be harnessed to political agendas, the means by which they could challenge existing world views and, finally, illustrate how they could be completely marginalised.

CHAPTER 10

A Frozen History of the Past: Antarctica, Gondwana and an Unfulfilled Dream

By the end of his career Du Toit noted that he had travelled to most of the major landmasses that comprised Gondwana—Africa, Australia, India and South America—and he had built intellectual relationships with like-minded scientists in these regions. He also had a more than passing familiarity with the geology of North America and Europe. The glaring omission was Antarctica. Yet it was the fossilised botanical material of Antarctica that provided compelling evidence for continental drift.

Early exploration of the Antarctic was driven by Europeans hunting whales and seals, following in the footsteps of the intrepid James Cook who, dissuaded by the barrier of the Antarctic Circumpolar Current, failed to reach land. Cook's negative portrayal of the icy, desolate region did little to prevent subsequent exploration from the early nineteenth century. It was on these hunting expeditions that the first fossil discoveries of ancient flora were made in the Antarctic. The evidence of plant remains in what was viewed as a barren wasteland indicated that the climate of Antarctica was vastly different in the past. An American seal-hunting expedition, the United States Exploring Expedition, carried on board a naturalist, James Eights, who paid particular attention to the geology of the South Shetland Islands that contained remnants of carbonised wood. Subsequent expeditions between 1830 and 1843 reported further fossil finds. These discoveries were, however, given scant attention until the late

S. Chetty, *'Africa Forms the Key'*, Cambridge Imperial and Post-Colonial Studies, https://doi.org/10.1007/978-3-030-52711-2_10

nineteenth century when discoveries much further north galvanised the embryonic field of palaeobotany.[1]

The discovery of fossilised flora in the Arctic which suggested a far more temperate climate in the past led to the proposition that it was, in fact, the Arctic that was the source of flora which had then moved south—even fossils found in the southern hemisphere were ascribed northern origins. This paved the way for greater investigation of the palaeobotany of the Antarctic. A Norwegian hunting expedition led by Captain C.A. Larsen took on a more scientific aspect when Larsen and his men made a large fossil find on Seymour Island in 1892. They were met by the Dundee Antarctic Expedition out of Scotland and some of the fossils were traded for tobacco, resulting in this material finding its way to Scotland where they were the subject of study with the conclusion that they bore a resemblance to their counterparts in the north. Larsen returned to Seymour Island two years later, once again highlighting the site's importance for fossilised material. He was also placed in charge of the Swedish South Polar Expedition. Despite numerous setbacks culminating in a shipwreck that forced the men to winter in the Antarctic, the Swedish South Polar Expedition was arguably the most successful thus far, finding flora that dated from the Jurassic and Cretaceous periods as well as fossilised wood. This was overwhelming proof that Antarctica (like the Arctic) had had a very different climate in the past. The next notable expedition was the British National Antarctic Expedition between 1901 and 1904 led by Captain Robert Falconer Scott which included the geologist, Hartley Ferrer. The importance of the fossilised plant material obtained on this expedition was only realised two decades later when remnants of *Glossopteris*, dating from the Permian, were found. This late realisation meant that the glory for the *Glossopteris* discovery fell on Scott's final expedition.[2] A final noteworthy expedition is that undertaken by Ernest Shackleton (who had also formed part of Scott's earlier expedition). At the end of 1908, Shackleton and his men ascended the Beardmore Glacier to find massive coal deposits, fossilised wood and the imprints of ancient leaves in sandstone. Shackleton's expedition became the first to obtain

[1] David J. Cantrill and Imogen Poole, *The Vegetation of Antarctica through Geological Time* (Cambridge, Cambridge University Press, 2012), pp. 2–4.

fossil evidence from the Antarctic continent—earlier samples had been obtained from surrounding islands.[3]

In February 1912, two months after Roald Amundsen had beaten him to the South Pole, Robert Falconer Scott halted atop the Beardmore Glacier in the Antarctic in order to go "geologising". The result led to the collection of almost 16 kilograms of rock samples for an already overburdened and exhausted expedition team. This was a small part of the 40,000 samples amassed by the Scott expedition but, arguably, the most important for the proponents of continental drift theory.[4] Housed in the Natural History Museum in London are the fossilised samples found with Scott's frozen body eight months after his death. Along with the rocks of the Beardmore Glacier was a fossilised specimen of *Glossopteris*.[5] This specimen would acquire an importance that was out of all proportion to its size.[6]

Antarctica is an old continent—approximately 3.8 billion years old—and has been part of a number of supercontinents, with Gondwana only the latest incarnation. The remnants of other supercontinents such as Rodinia and Pangaea can be found in landmasses in both the northern and southern hemispheres and reflect, in the fossil record, their association with the now isolated Antarctica. The Transantarctic Mountains contain coal deposits a quarter of a billion years old—the remnants of trees—that indicate an amenable climate in the past. There is fossil evidence of dinosaurs and ancient mammals and birds and, of course, flora. Much of what is known of the natural history of Antarctica, however, comes from the rock and fossil samples that are easily obtainable—the coastlines and mountains. The rest remains entombed in the ice.[7]

More than 600 million years ago, Antarctica formed part of an ancient supercontinent known as Rodinia. As Rodinia began to break up, fragments drifted southwards to form Gondwana—South America, Australia,

[2] Cantrill and Poole, *The Vegetation of Antarctica*, pp. 4–7, 8.

[3] Jeffrey D Stilwell and John A Long, *Frozen in time: Prehistorical Life in Antarctica* (Collingwood, Csiro Publishing, 2011), pp. 32–33.

[4] Robin McKie, "Scott of the Antarctic: the lies that doomed his race to the pole" in *The Guardian*, Saturday, 24 September, 2011, http://www.theguardian.com/uk/2011/sep/24/scott-antarctic-lies-race-pole, Accessed 17 October 2016.

[5] Ted Nield, *Supercontinent: Ten Billion Years in the Life of Our Planet* (London, Granta Books, 2008), p. 67.

[6] McKie, "Scott of the Antarctic".

[7] Stilwell and Long, *Frozen in Time*, pp. 1–3.

Africa and Antarctica. A hundred million years later, Gondwana was complete and would remain that way for another 300 million years. At this time Gondwana and Laurasia came together to form a single massive landmass, Pangaea. This was a period of extreme cold, as the Carboniferous Period came to an end, and extensive glaciation occurred. The Carboniferous gave way to the Permian Period which was more temperate and, as the glaciers retreated, they left behind the detritus which would become rich fossil troves. It is during the Permian Period that we find *Glossopteris*.[8]

The appellation *Glossopteris* was given to a fossilised leaf by Adolphe Theodore Brongniart in 1828. Brongniart, a geologist, mistakenly believed that the leaf was a remnant of a fern, an ancient relative of the Hart's Tongue Fern and *Glossopteris* thus means "tongue fern".[9] This was a most pervasive plant, flourishing in the Permian climate. While the plants resembled ferns, unlike ferns—which propagate via spores—*Glossopteris* had seeds. A tree, standing as much as 20 metres tall, *Glossopteris* had large leaves ranging from 10 centimetres to as much a metre, which it shed annually. There were a large number of species that spread over all the Gondwana-associated landmasses. As a testimony to its success, its fossil remains in Antarctica occur in most of the Permian rocks and are preserved in very fine detail.[10]

As a supercontinent, Gondwana's interior—distant from the temperate influence of the ocean—was subject to climactic extremes. Due to the proximity of Gondwana to the South Pole, sunlight was at a premium and two distinct seasons were evident—one of total darkness and the other of dim light. The plant took advantage of the weak sunlight and the summer months were a period of tremendous growth. Those trees closest to the Tethys Ocean were subject to tremendous monsoon rains on a scale far great than today that also had an adverse effect on the available sunlight. Despite these challenges, to put it mildly, *Glossopteris* was prolific. Dense forests of *Glossopteris* trees could be found in the coastal areas, with as many as a thousand trees per acre. With the end of the growing season came the perennial darkness of winter and leaves were shed in thick layers. Despite the temperate climate of the Permian—that can be contrasted

[8] Stilwell and Long, *Frozen in Time*, pp. 11, 81–83, 85.

[9] Nield, *Supercontinent*, p. 67.

[10] Stilwell and Long, *Frozen in Time*, p. 85.

with the vast forests of the Carboniferous that led to the accumulation of most of the world's coal deposits—the availability of water in the coastal regions of Gondwana promoted coal formation. Therefore, despite going extinct hundreds of millions of years ago, *Glossopteris* endured in the present in the form of coal deposits and as fossils. This distinctive plant thus became an important means of proving the links between the disparate landmasses that once formed Gondwana.[11]

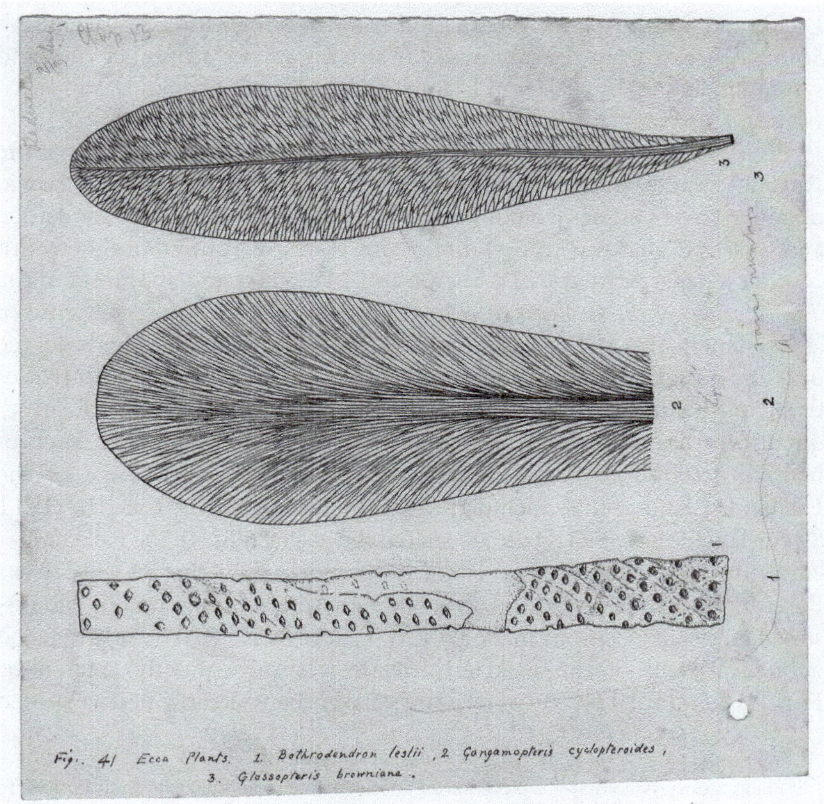

Sketch of Ecca plants with *Glossopteris browniana* at top (University of Cape Town Libraries, mss_bc722_c2_8_001)

[11] Nield, *Supercontinent*, pp. 67, 120–122.

In addition to the fossil evidence, a major area of comparison in Du Toit's *A Geological Comparison of South America with South Africa* was the Falkland Islands in terms of their rock composition and stratigraphy. Du Toit later included South Georgia as well in his reconstruction of Gondwana, making the connection between the Antarctic and South America. While Du Toit's reconstruction was remarkably visionary he was hampered by the lack of adequate knowledge of these South Atlantic islands.[12] The Scotia Arc lies largely on the junction between the South Pacific and South Atlantic Oceans and comprises the Scotia Sea separating South America from the Antarctic. It also comprises a number of island systems including the Falkland Islands, South Georgia and South Sandwich Islands.[13]

Charles Darwin was the first to give an account of the geology of the Falkland Islands whilst sailing on the *Beagle* in 1833 and 1834. Darwin obtained fossil samples from mudstone and also highlighted the sandstone and quartzite that was to be found there. As early as the mid-nineteenth century a paper written by D. Sharpe and J.W. Salter and appearing in the *Transactions of the Geological Society of London* noted with some sense of astonishment that the palaeontological evidence suggested a high degree of correlation between the formations of the Falkland Islands with that of their South African counterparts.[14] It was, however, as a result of subsequent expeditions more than 70 years later that there was a more comprehensive description of the fossilised Devonian flora and fauna of the Falkland Islands—that included the ubiquitous *Glossopteris*. In 1911 Swedish geologist, T.G. Halle associated the glacial tillite of the Falklands—currently named the Fitzroy Tillite Formation—with that of the Dwyka Formation in South Africa. Halle's findings were substantiated by the fossilised faunal associations made by J.M. Clarke in 1913. The first detailed geological map of the Falkland Islands was subsequently created by H.A. Baker in 1924. By this point, Baker had already been exposed

[12] Philip Stone, "Geological exploration of South Atlantic islands and its contributions to the continental drift debate of the early 20th century" in *Proceedings of the Geologists' Association*, 2015, Vol. 126, 266–281, p. 1. Taken from http://nora.nerc.ac.uk/510724/1/SouthAtlanticIslands-pga.pdf, Accessed 12 October 2016.

[13] Andres Maldonada, Ian W.D. Dalziel and Philip T. Leat, Corrigendum to "The global relevance of the Scotia Arc: An introduction" in *global and Planetary Change*, vol. 133, October 2015, p. 378. Taken from http://www.sciencedirect.com/science/article/pii/S0921818114001404, Accessed 6 December 2016.

[14] Stone, "Geological exploration of South Atlantic islands", pp. 4, 25.

to—and influenced by—Wegener's work and mapped the geology of the Falkland Islands in light of continental drift.[15]

It was, however, neither Wegener nor Du Toit who were the first to make the connection between the Scotia Arc, South Africa and South America. In 1821, a Russian explorer, Thaddeus von Bellingshausen had suggested that South Georgia and the South Sandwich Islands were the visible indicators of an underwater mountain system that linked the Scotia Arc with South America.[16] Fifty years later, Austrian Heinrich Klutschak stated in "A Visit to South Georgia":

> [a]s regards its relation to other continents, South Georgia is part of a submarine range of high mountains connecting the uplands of South America with the Antarctic continent. The highest peaks of this range between Cape Horn and the Cape of Good Hope are represented by the Falkland Islands, South Georgia, the South Orkneys [and] the South Sandwich Islands. With the exception of the Falkland Islands...they are all of the same volcanic nature.[17]

In the first decade of the twentieth century, Eduard Suess focused on the atypical nature of the geology of the Falkland Islands that appeared to have a greater congruence with the geology of South Africa than it did with the geology of South America. By 1927, then, when Du Toit made his argument, using the evidence of these South Atlantic islands, he was able to build upon the foundations laid by early explorers, palaeontologists and geologists. Du Toit, was the first to reconstruct the early movement of the Falkland Islands, rather than simply viewing them in terms of their contemporary proximity to South America—which could be used by drift detractors to support the notion of "land bridges". The Falkland Islands are presently situated less than five hundred kilometres off the coast of South America and this close proximity did not therefore preclude the notion of a land bridge connecting the two that could have been submerged by rising sea levels in the vein of the Bering Strait and the English Channel. To demonstrate what he perceived to be the actual movement of continents—which was counterintuitive to those who supported the land

[15] Stone, "Geological exploration of South Atlantic islands", pp. 4, 20, 22.

[16] Stone, "Geological exploration of South Atlantic islands", p. 8.

[17] H.W. Klutschak, "Ein Besuch auf Sud Georgien", R.S. Boumphrey, translator, in *British Antarctic Survey Bulletin*, 12, 1881. Quote taken from Stone, "Geological exploration of South Atlantic islands", pp. 47–48.

bridge hypothesis—Du Toit suggested instead that the Falkland Islands were initially further north, forming the connection between the Ventania region of Argentina and the Cape Province in South Africa.[18]

It is unsurprising then that Du Toit expressed a special interest in a South African research expedition to Antarctica. There had been a historic link between the expeditions to Antarctica and South Africa. Some of the expedition members who had accompanied both Shackleton and Scott had had an association with South Africa, much of it dating to the South African War at the turn of the century. Frank Wild had enlisted in the Royal Navy during the conflict and was stationed in Simon's Town in 1901 when he volunteered for the National Antarctic Expedition under Scott. Chosen from thousands, Wild took part in that expedition and in a subsequent expedition in 1907. He also served with Shackleton, taking command of the latter's expedition after his untimely death. Wild's ashes were later reinterred in South Georgia in proximity to those of Shackleton. Ernest Joyce was also a member of the Royal Navy during the South African War and, like Wild, signed up for Scott's expedition in Simon's Town. He subsequently took part in a number of Antarctic expeditions, playing a significant role in taking care of the dogs, sleds and establishing supply stations along the expedition routes. Despite a history of poor health, Lawrence Oates served in the 6th Inniskilling Dragoon Guards during the South African War and was shot in the leg by Boer fire near the town of Aberdeen and later suffered a second wound when he returned to duty. While never fit enough to join Scott's expedition as an official member, Oates donated £1000 towards the expedition in 1910 and served in an unofficial capacity, taking care of the horses. Suffering from scurvy and afflicted with gangrene in his weakened leg, he sacrificed himself in a blizzard in an effort not to impede the remainder of the expedition.[19]

As with these early explorers, Antarctica exerted a powerful hold on Du Toit. His papers include a collection of newspaper clippings detailing exploration in Antarctica prior to the outbreak of the Second World War. Of particular interest to him were articles detailing Antarctic topography as well as fossil finds and, in some instances, the connections between continents were made explicit. These recent explorations, however, differed from older expeditions due to the use of the new technology

[18] Stone, "Geological exploration of South Atlantic islands", pp. 8, 9.

[19] Dr Sydney Cullis, "The Heroic Antarctic Explorers and the South African War of 1899–1902" in *The South African Military History Society* Newsletter No 456, August 2017, pp. 1–4.

available. A five-month-long exploration undertaken by Lincoln Ellsworth in 1935 with the aim of flying over the South Pole concluded with Ellsworth's belief that "the mountain ranges of Antarctica belong to the Andes system of South America".[20]

The British Grahamland expedition, undertaken two years later, substantiated the comparisons through an investigation of the stratigraphy. Researchers on the expedition reported that the igneous rock of Grahamland resembled that of the Andes and that the fossilised fauna and flora discovered in the sedimentary rock comprising the Alexander I Longshore suggested a far more temperate climate evident in Western Antarctica in the past.[21] Between 1933 and 1935, American Admiral Richard E. Byrd carried out an expedition in Antarctica with a pioneering use of mechanised transport. Scientific research formed a strong component of the expedition with geological surveys carried out, a number of Antarctic mountains explored and magnetic and seismic measurements taken.[22] An article detailing this expedition contained Du Toit's handwritten interventions with descriptions of the sea and mountain chains underlined in red accompanied by his speculation that the latter possibly signified the "N. side of Ross fault?"[23]

Yet this evidence was incomplete and second hand and must have proved both tantalising and frustrating for Du Toit. A solution suggested itself six years later, in the midst of the international upheaval that was the Second World War. Fellow geologist, Lester King of Natal University College proposed to Du Toit the possibility of a South African-led expedition to the Antarctic comprising geologists *au fait* with African geology and with the intent of drawing geological comparisons between Africa, Madagascar and Antarctica.[24] Du Toit's response was enthusiastic albeit with some reservation (which would prove to be founded) regarding

[20] UCT-JL: Alex L. Du Toit Papers—BC 722: J-Other, J1—South African Antarctic Expedition, "Andes Continue to Antarctica: Mr Lincoln Ellsworth's Contention" in The Star, 4 February 1935.

[21] UCT-JL: Alex L. Du Toit Papers—BC 722: J-Other, J1—South African Antarctic Expedition, "Discoveries Unexpected and Important—'Archipelago' really part of Antarctic continent, Fossil Invertebrates Found" in The Star, 25 August 1937.

[22] "Richard E. Byrd: Byrd Antarctic Expedition II" http://www.south-pole.com/p0000108.htm, Accessed 5 March 2018.

[23] UCT-JL: Alex L. Du Toit Papers—BC 722: J-Other, J1—South African Antarctic Expedition, Article from 19 November 193(9?).

[24] UCT-JL: Alex L. Du Toit Papers—BC 722: J-Other, J1—South African Antarctic Expedition. Letter by geologist LC King (Natal University College) to Du Toit, 25 October 1943.

South Africa's ability to mount such an expedition. Fully supporting King's suggestion, he proposed the possibility of Norway being involved and the use of helicopters to facilitate exploration and research with a focus on the Wedell Sea. For Du Toit, the proposed Antarctic Expedition had the potential to greatly substantiate his hypothesis, providing a major piece of the puzzle: "Having nailed my colours to the Pole, I am in duty bound to the testing out of this keystone of my hypothesis. If it breaks down there, then I am afraid c. rift is just moonshine. At present I fail to see any flaw in the reasoning that the remainder of the Transkei must form part of the Antarctic mass".[25] Now 65 years old, Du Toit was also aware that his participation in the expedition would be at a distance and he wryly indicated to King that he would be content with watching the expedition on film and examining the rock specimens obtained.[26]

With Du Toit's blessing and unqualified support, King was ready to proceed. He approached the Geological Society of South Africa (GSSA) with his plan for a South African Antarctic Expedition that would pay particular attention to the accumulation of data related to geology, geography and meteorology. Moreover, as E. Mendelssohn, secretary of the GSSA, pointed out, the scientific aspect of the expedition could be harnessed to national interests as well and the expedition was approved.[27]

A detailed memorandum was subsequently drawn up detailing the aims of the proposed expedition. Broad objectives included a survey of the topography for geographical interest, measurements of ice and magnetism for the purposes of geophysics and various atmospheric measurements taken from a subsidiary base on Bouvet Island in the belief that the weather of the southern hemisphere was largely related to conditions at the pole. It was, however, in the field of geology that the expedition was expected to make its greatest contribution by buttressing the work of Du Toit. The memorandum highlighted Du Toit's contribution as a South African to continental drift theory which was believed to be as revolutionary for geology as Darwin's theory of evolution had been for biology. Du Toit

[25] UCT-JL: Alex L. Du Toit Papers—BC 722: J-Other, J1—South African Antarctic Expedition. Letter by Alex Du Toit to LC King, 17 November 1943.

[26] Letter by Alex Du Toit to LC King, 17 November 1943.

[27] UCT-JL: Alex L. Du Toit Papers—BC 722: J-Other, J1—South African Antarctic Expedition. Letter by E. Mendelssohn, secretary of the Geological Society of South Africa to Alex Du Toit on 22 May 1944.

was credited with being a key figure in generating acceptance for the former connection between South America and South Africa. The geological aim of the explanation was therefore to find correlations between the stratigraphy of the Wedell Sea area of Antarctica and South Africa as reconstructed by Du Toit.[28] Du Toit himself subsequently submitted a memorandum to King detailing the specific geological aims of the expedition that focused on the evidence utilised by Du Toit—glaciation, volcanism, stratigraphy and chemical composition of rocks and minerals:

Memorandum by Alex Du Toit to LC King

(1) "General knowledge which would naturally be new.
(2) Information regarding the Archaean Complex, & particularly the dominant trends of its foliation
(3) The discovery of deposits, like those of Madagascar & Ceylon, of graphite, sillimanite rocks, & rare minerals such as corundum, monazite, thorianite, etc
(4) The presence of fossiliferous palaeozoics. A lump of Lower Cambrian Archaeocyathus limestone was dredged from the Weddell Sea.
(5) The existence of any glacials at the base of the Gondwana Beds, or indeed of any signs of glaciation prior to the Pleistocene. At present the lack of anything positive from this 'polar' continent has proved most disconcerting to geologists & meteorologists alike.
(6) The continuation of the Cape Permo-Triassic folding deduced from theoretical considerations
(7) The presence or absence of Mesozoic eruptives. By analogy the dolerites should become scarcer on nearing that fold belt
(8) The existence of Cretaceous & Tertiary marine strata, the climatic conditions under which they were deposited & the affinities of their faunas & floras.
(9) The tracing of the Andean Fold-belt southwards and westwards from Hearst Island.
(10) The possible occurrence of young volcanos [*sic*] & the nature of their lavas".[29]

[28] UCT-JL: Alex L. Du Toit Papers—BC 722: J-Other, J1—South African Antarctic Expedition. The Geological Society of South Africa, "Memorandum upon a Proposed South African Antarctic Expedition".

[29] UCT-JL: Alex L. Du Toit Papers—BC 722: J-Other, J1—South African Antarctic Expedition. Memorandum by Alex Du Toit to LC King, 20 November 1944.

And Du Toit was literally prepared to back the expedition by any means at his disposal, donating £50 "as a token of my immense interest" to the Antarctic Exploration Committee as they attempted to garner financial and government support for the endeavour.[30]

Yet the Committee faced an uphill battle. The initial memorandum—drawn up by King—was an ambitious plan requiring substantial resources. Envisaged as taking place when the War eventually concluded, the expedition could take two possible forms—a short trip that would focus on geological research and a longer and more rewarding trip that could potentially offer more data. The former, largely making use of surveys conducted by sea and air would at best be cursory and ultimately ineffective. The second option, however, required far greater expenditure in terms of time and money with reconnaissance flights to determine suitable bases and areas of exploration, the associated creation of fuelling sites for flights into a largely unknown interior and for the purposes of maintaining communication links, the provision of sleds and dogs to aid surveys on the ground and even the use of helicopters to search for suitable outcrops. Also needed would be a ship to transport men while being able to tackle the icy waters and the vast amount of supplies needed to maintain the expedition over an extended period. The projected cost was in the region of £80000.[31]

To raise the money for the expedition, the Committee hoped to benefit from the generosity of individuals and national scientific associations along with the participation of the government. An extensive role was envisaged for the South African government in terms of providing financial assistance, the instrumentation for meteorological observations and the use of aeroplanes. It could further request co-operation and assistance from other states such as extraneous war supplies that could be adapted for the expedition. There was even the possibility of co-operating with Norway—as suggested by Du Toit—which controlled areas of prospective exploration such as Crown Princess Martha Land and Bouvet Island. Drawing upon the conventions set by expeditions dating from the early twentieth century, money could also be raised through selling exclusive rights to

[30] UCT-JL: Alex L. Du Toit Papers—BC 722: J-Other, J1—South African Antarctic Expedition. Letter by Alex Du Toit to LC King, 17 May 1945.

[31] The Geological Society of South Africa, "Memorandum upon a Proposed South African Antarctic Expedition".

newspapers to publish articles on the expedition, the giving of public lectures and the publication of a book and film after the expedition. Another possibility was to approach international organisations such as the Carnegie Corporation (which had provided the funds for Du Toit's initial trip to South America in 1923).[32] All avenues were to be explored and, as King rather wryly stated to Du Toit, "If possible, I propose to lay under contribution even the small pennies of South African school children".[33] Another tongue-in-cheek proposal was the promise of posterity: "Do you think we could promise to name peaks after people for £1000 and mountain ranges for £10,000? or is the Antarctic terrain not yet up for auction?"[34]

There was also the proposal for the formation of a Committee—the South African Antarctic Research Committee—with a membership comprising the various interested figures—scientists, government and financial and technical advisors. There was the hope of requesting Jan Smuts to become the "patron" of the expedition.[35] This was an idea with which Du Toit was in full agreement, "It was a fine move to rope in F-M [Field Marshal] Smuts as patron, a job we shall all hope he accepts".[36]

The memorandum concluded with the assertion that the envisaged expense and resources required for the expedition were considered negligible when it was weighed against an affirmation of South African prominence in the southern hemisphere, "if South Africa is to remain a significant nation in the Southern Hemisphere, it will do well not to neglect the study of what lies south of it as north". Parallels were drawn with South Africa's fellow Commonwealth member (and rival), Australia, which had done the same and so maintained a dominant presence in the Southern Ocean.[37]

The Committee members began with a sense of optimism and enthusiasm, approaching Smuts soon after the decision was taken to request his

[32] The Geological Society of South Africa, "Memorandum upon a Proposed South African Antarctic Expedition".

[33] UCT-JL: Alex L. Du Toit Papers—BC 722: J-Other, J1—South African Antarctic Expedition. Letter by LC King (Natal University College) to Du Toit, 11 September 1944.

[34] UCT-JL: Alex L. Du Toit Papers—BC 722: B3 Correspondence: B3.3 Correspondence (c.1937–1947). Letter by LC King (Natal University College) to Du Toit, 2 May 1944.

[35] The Geological Society of South Africa, "Memorandum upon a Proposed South African Antarctic Expedition".

[36] UCT-JL: Alex L. Du Toit Papers—BC 722: J-Other, J1—South African Antarctic Expedition. Letter by Alex Du Toit to LC King, 31 October 1944.

[37] The Geological Society of South Africa, "Memorandum upon a Proposed South African Antarctic Expedition".

involvement. Smuts, on the other hand—despite his lifelong association with scientific endeavour in South Africa—was non-committal. He turned down the offer to serve as patron of the expedition but agreed to a meeting of selected Committee members that included King. Smuts, however, was careful not to make definite promises. Distracted by the war—which was still ongoing—he considered the expedition promising but felt that the Committee should focus on conducting further investigation into making it financially feasible and in forming links with possible supporters. He literally adopted a "wait-and-see" approach, merely "[stating] that he would be prepared to give further consideration to the matter".[38] Smuts's obvious lack of enthusiasm can be contextualised by a war that was stretching South African resources overseas while fomenting domestic turmoil among South Africans either ambivalent or actively hostile towards South Africa's participation in the conflict. Yet his response must have been a source of disappointment to the Committee and set the trend for what was to follow.

In addition to Smuts's lacklustre support—which was less than the Committee had hoped for—was the problem of funding. A letter from the Scott Polar Research Institute based in Cambridge, England, politely rebuffed the Committee's request for funding citing their own financial straits and reliance upon volunteerism. The Institute further excluded the possibility of obtaining funding from their affiliates and associates that would be unable to provide the necessary support for such an ambitious undertaking.[39] Subsequent correspondence from the "Discovery" Committee based in the Colonial Office in London—and named for the Discovery expedition in 1901 that included Scott as well as Shackleton—also indicated that they had been approached by the South Africans. While making little reference to funding the expedition, the "Discovery" Committee was more inclined to offer a ray of hope, suggesting the possible assistance of their ship, the Discovery II, to aid their expedition's efforts when the war had concluded. They also offered some unsolicited advice by suggesting that the Committee place meteorological data on

[38] UCT-JL: Alex L. Du Toit Papers—BC 722: J-Other, J1—South African Antarctic Expedition. Minutes of South African Antarctic Research Committee Meeting under the Geological Society of South Africa, held in Johannesburg, 15th November 1944.

[39] UCT-JL: Alex L. Du Toit Papers—BC 722: J-Other, J1—South African Antarctic Expedition. Letter from the Scott Polar Research Institute, Cambridge England to SA Antarctic Research Committee, 21 November 1944.

equal footing with geological research as the former would potentially be of great significance. To this end, there should be greater emphasis on atmospheric measurements.[40]

Undeterred by the difficulties in raising the necessary funds for the expedition, the expedition received keen interest from the wider South African scientific community. This was enhanced by first-hand testimony. Professor R.W. James who had been on Ernest Shackleton's expedition aboard the "Endurance"—which had subsequently been destroyed by the ice—gave a lecture to the Royal Society of South Africa in October, 1945. James's experience provided a cautionary tale of the need for awareness of ice conditions that could adversely affect a potential expedition and the possibilities of research. Yet these possibilities were significant. Atmospheric measurements offered the potential to understand the extremes of South African climate with its associated agricultural and economic effects. There was an additional need for groundwork as aerial reconnaissance alone was insufficient in clearly distinguishing landmarks—both forms were necessary for an accurate mapping of the frozen continent. Fieldwork also allowed for solving what were currently intractable "geological problems" and specific reference was made to the Antarctic's more temperate climate in the past and Du Toit's hypothesis of the glaciation that originated in South Africa. This would allow for greater understanding of climatic cycles—and would lend support to Du Toit's work.[41]

The press adopted the project with enthusiasm, highlighting the importance of its research potential in terms of the meteorology but with an added economic emphasis. The *Cape Times* was particularly taken with Scott's discovery of coal on the southern continent as well as the possibility of other minerals—an emphasis related to the specific mining economy of South Africa. The article emphasised South Africa's suitability for Antarctic exploration due to the country's expertise in mining and mineral exploration. Underlying this lurked the thrill of discovery, of rendering the great unknown known. The article's support of the expedition also had an ideological bent, ostensibly criticising South Africa's hitherto largely "passive" role in Antarctic exploration where the country merely

[40] UCT-JL: Alex L. Du Toit Papers—BC 722: J-Other, J1—South African Antarctic Expedition.—Letter from the Secretary of the "Discovery" Committee, the Colonial Office, London to the SA Antarctic Research Committee, February 1945.

[41] UCT-JL: Alex L. Du Toit Papers—BC 722: J-Other, J1—South African Antarctic Expedition.—"Antarctic Exploration" in *Cape Times*, 25 June 1945.

played the role of "interested spectator" rather than active participant. The proposed Antarctic Expedition offered the possibility of enhancing South Africa's international reputation and was thus a means of asserting the country's prowess "in the eyes of the rest of the world". With the benefits of South African Antarctic exploration clearly spelled out, the article was unambiguous in its support for the proposal, expecting little different from the government and the public at large in terms of the financial cost.[42]

A subsequent article continued in this manner, seizing on the South African expedition with gusto, considering it as "by far the biggest scientific expedition ever organised" in the country and describing it in the language of war as a "scientific blitz" that would be unleashed onto the Antarctic. In addition to the prospective mineral resources and benefits for an understanding of climate and even the migration of fish, the article highlighted the dominion of science over nature. Not only would the proposed expedition divulge the mysteries of the last unexplored continent, its extreme environment could also be harnessed to the benefit of mankind. The article—with an almost science fiction bent—predicted the energy derived from the intense blizzards characterised by wind speeds of almost five hundred kilometres an hour could be used to power the mining industry. The ebullience of the article was evident in its observation that there was also underway the design of a crest for the expedition and there was little hint of Smuts's reluctance. On the contrary, the Prime Minister was named as actively "considering in what way the Government can assist the expedition" that would likely get underway at the end of 1947. Financial obstacles were airily glossed over and the necessary equipment was believed to be easily obtainable from the government and academic institutions.[43] Yet, contrary to the optimism displayed by the *Cape Times*, South African involvement was already limited by circumstances beyond the control of the GSSA.

The prohibitive costs and unforeseen difficulty in obtaining support led to a re-envisioning of the expedition with Smuts at the forefront. With South Africa unwilling to shoulder the burden alone, Smuts was more

[42] UCT-JL: Alex L. Du Toit Papers—BC 722: J-Other, J1—South African Antarctic Expedition.—"Eyes to the South" in *Cape Times*, 1 October 1945.

[43] UCT-JL: Alex L. Du Toit Papers—BC 722: J-Other, J1—South African Antarctic Expedition.—"S. African Expedition to Antarctica" in *Cape Times*, 11 October 1945.

amenable to an "International Expedition" where the costs—and risk—could be shared. To this end, the Prime Minister instructed physicist, Dr Basil Schonland, chosen by Smuts as the first president of the Council of Scientific and Industrial Research established in 1945[44]—to broach the proposed expedition at the upcoming Commonwealth Scientific Conference and during the course of his later visit to the United States. Smuts suggested the idea of an international expeditionary effort as part of what he envisaged as a "Polar Year" that would prioritise international cooperation and allow for the use of technology developed during the course of the war which would ultimately necessitate the expedition taking place later than originally planned. For Smuts, the Committee's role would simply be to continue in their efforts to accumulate data regarding the requirements of the expedition and their hopes for Antarctic research would best be served through co-operation with Schonland. With little choice, the Committee signalled their agreement, holding on to the tenuous belief that should the "Polar Year" not bear fruit, the Smuts-led government would still be willing to back a South African effort.[45]

Schonland complied with Smuts's request in an address to the Commonwealth Scientific Conference where he spoke of the South African government's unwillingness to "go it alone", putting forward the GSSA's proposal for "an international expedition shared by countries interested in the Antarctic" that could lead to the setting up of various research bases and allow for extensive scientific study. The South African proposal was not the only one of its kind and a proposal was also presented for an Anglo-Norwegian expedition with a specific focus on meteorological research.[46] This expedition would focus on the Norwegian-controlled territory of New Schwabenland that had the potential to offer great insight into glaciation and the "climatic history of the Antarctic".[47] It was on this latter expedition that the South Africans would ultimately pin their hopes

[44] "Basil Schonland (1896–1972)" http://www.thepresidency.gov.za/national-orders/recipient/basil-schonland-1896-1972, Accessed 5 March 2018.

[45] UCT-JL: Alex L. Du Toit Papers—BC 722: J-Other, J1—South African Antarctic Expedition.—Minutes of South African Antarctic Research Committee Meeting under the Geological Society of South Africa, held in Johannesburg, 19th November 1945.

[46] UCT-JL: Alex L. Du Toit Papers—BC 722: J-Other, J1—South African Antarctic Expedition.—"S.A. May Take Part in Antarctic Expedition" in *Rand Daily Mail*, 29 June 1946.

[47] UCT-JL: Alex L. Du Toit Papers—BC 722: J-Other, J1—South African Antarctic Expedition.—"Scientific Expedition to Antarctic Proposed" in *Rand Daily Mail*, 30 June 1946.

as secondary participants. This was also reflected in a greater emphasis on climate rather than geological research as had initially been envisaged by the GSSA, as well as a shift to New Schwabenland as the focus of the expedition rather than the Wedell Sea located in the Southern Ocean as proposed by Du Toit.

A memorandum drawn up by the GSSA two years later suggested that the plan to adopt a "Polar Year" at the expense of a South African-led expedition to the Antarctic was a *fait accompli*. The GSSA thus turned its attention to the proposed research plan within this new context. The emphasis was on international co-operation with the establishment of various research stations engaged in scientific research that would comprise interested parties. This included the southern states that were in close proximity to the Antarctic—South Africa, Australia, New Zealand, Argentina and Chile—and northern states such as Britain, Norway and the United States. This allowed for more options in terms of funding from both governments and private institutions with the possibility of research getting underway from late 1947. International co-operation also allowed for multiple expeditions and increased the area that could be opened to research and the setting up of research stations. Findings would be shared and the spirit of co-operation was a refreshing change from the divisive conflict that had been ongoing for almost six years. The research agenda also proved to be more ambitious with the inclusion of fields such as biology, physics and geography. In the field of geophysics, the emphasis was on determining the dimensions of the polar ice cap and the taking of gravity and electrical resistance measurements. In addition to the study of rock outcrops and above ground survey, envisaged geological research had a particular focus on continental drift: "Examination and collection in the Weddell Sea region to be with special reference to the theory of Continental Drift".[48]

By early 1946, then, the GSSA was still optimistic and had come to terms with a more co-operative effort that would, nonetheless, allow for research to be carried out in support of Du Toit's drift theory. However, they would have to contend with the disappointing reality in September of that year. A meeting of the South African Antarctic Research Committee was held in Johannesburg on the 13th of September. Du Toit was unable

[48] UCT-JL: Alex L. Du Toit Papers—BC 722: J-Other, J1—South African Antarctic Expedition.—"The Geological Society of Society Africa Memorandum upon a Proposed 'International Polar Year' in the Southern Hemisphere", Johannesburg, 30th November 1946.

to attend the meeting but sent his apologies the day before as well as a recommendation where he "strongly [suggested] that society freely offer fullest support information and financial aid to proposed Anglo-Norwegian expedition in return for some actual share therein".[49] The meeting was an exercise in making the best of a difficult situation with suggestions that South Africans be allowed to take part in other international expeditions including a planned Anglo-Scandinavian excursion to New Schwabenland. A possibility was raised of getting permission to have young South African scientists trained on the Falkland Islands in order to prepare them for future expeditions and research at Antarctic bases. Despite these alternate proposals, the conclusion of the meeting suggested a tone of despondence, "Plans for a purely South African expedition to the Antarctic to be held in abeyance".[50] By February 1947, then, the writing was on the wall and, far from being at the forefront of the most ambitious South African scientific undertaking, an item on the agenda of the Committee dealt with a proposed trip to the Antarctic and was non-committal: "Possible South African Participation in the New Schwabenland Expedition".[51] A lengthy process of expectation and negotiation had come to naught and Alex Du Toit's hopes had ultimately been dashed.

Yet Du Toit had continued to maintain an interest in Antarctic exploration with the possibility of South African involvement—albeit at a far greater remove than he had initially hoped. In the year preceding his death, Du Toit paid particular attention to an expedition led by Norwegian Major-General Hjalmar Riiser Larsen. Contextualised in the new spirit of co-operation heralded by the United Nations, the expedition would feature the combined efforts of Norway, Sweden and Britain. Larsen would simultaneously maintain contact with a current Australian expedition and the Antarctic Expedition—estimated to cost £100000—would be synchronised with a Danish meteorologically focused expedition in Greenland

[49] UCT-JL: Alex L. Du Toit Papers—BC 722: J-Other, J1—South African Antarctic Expedition.—"Minutes of South African Antarctic Research Committee Meeting under the Geological Society of South Africa", held in Johannesburg, 13th September 1946.

[50] "Minutes of South African Antarctic Research Committee Meeting under the Geological Society of South Africa".

[51] UCT-JL: Alex L. Du Toit Papers—BC 722: J-Other, J1—South African Antarctic Expedition.—Document detailing of meeting of the South African Antarctic Research Committee by The Geological Society of South Africa, Johannesburg, 6th February 1947.

with the ultimate aim of setting up research bases running the breadth of the globe from the Antarctic to the Arctic.[52]

Larsen, a seasoned veteran of the ice, who had taken part in Roald Amundsen's expeditions in the Arctic—arrived in Cape Town in June 1947 with the aim of getting South African involvement in the expedition. This would be through the Union Meteorological Service which had a base on the isolated and distant Tristan de Cunha that was expected to play a role in the expedition. While Larsen did not rule out more direct South African participation, this was unlikely to have occurred. Furthermore, the focus of the expedition would be meteorological with no mention of geological research and the key aim was addressing the rising global temperatures of the present rather than the glacial conditions of the past.[53]

The sheer cost and scope of Larsen's expedition suggested that the South African plan was too ambitious—despite its ideological aim of asserting national scientific endeavour. The context of the Second World War and its immediate aftermath, compounded by Smuts's international—and domestic—concerns meant that insufficient government support was forthcoming. Political and economic considerations took priority. The conclusion of the Second World War had thrown into stark relief the divisions within South African society, leading to Smuts ultimately being ousted from power in 1948 with the rise of the conservative National Party under the leadership of D.F. Malan and, with it, the establishment of the apartheid state. Du Toit would likely have been disappointed with the lack of government support for the Antarctic Expedition but the political climate had already been moving towards isolation and fragmentation where there was no place for an holistic view of the world.

[52] UCT-JL: Alex L. Du Toit Papers—BC 722: J-Other, J1—South African Antarctic Expedition.—"Joint Expedition to Antarctic: Combined Operation in Spirit of United Nations" in *Cape Argus*, 30 April 1947.

[53] UCT-JL: Alex L. Du Toit Papers—BC 722: J-Other, J1—South African Antarctic Expedition.—"Important South African Aid to Antarctic Expedition: Weather News from Lonely Island" in *Sunday Express*, 22 June 1947.

CHAPTER 11

The Final Years

On 13th December 1940, Du Toit penned his letter of resignation to De Beers due to a "desire to settle at the Cape and to devote some of my leisure to research work".[1] He was subsequently touched by De Beers' response which expressed regret at his resignation and granted him a further six months' leave of absence prior to his resignation with full pay of £150 per month. Du Toit, ever focused on geology, informed the mining powerhouse that a portion of the funds would be devoted to research.[2] Yet, even as Du Toit was eager to devote more time to his research, the world and its concerns were less inclined to leave him to his splendid isolation. His status as a geological expert and own sense of responsibility meant that he was constantly engaged in correspondence and played mentoring and advisory roles for young geologists. His diaries list correspondence with colleagues all over the globe and, as he records in the rough notes that appear to form the beginnings of his autobiography:

> I have been handicapped in numerous ways, by lack of particular opportunities, by restricted freedom of action, by the [nimbus?] of administrative duties—until I retired—but above all and down to the present moment, by

[1] UCT-JL: Alex L. Du Toit Papers—BC 722: A4—Personal Documents: Letter by Du Toit to De Beers, 13th December 1940.

[2] UCT-JL: Alex L. Du Toit Papers—BC 722: A4—Personal Documents: Letter by De Beers to Du Toit, 3rd February 1941 and Du Toit's response, 7th February 1941.

S. Chetty, *'Africa Forms the Key'*, Cambridge Imperial and Post-Colonial Studies, https://doi.org/10.1007/978-3-030-52711-2_11

> the time-consuming drudgery of correspondence—private and public, of business arising out of societies, council committees and such like. Despite all efforts the time available for pure research has grown steadily less, although suitable subjects have increased, and original ideas about them have not been less abundant or fertile ... Freed ... my output could easily have been doubled or even trebled.[3]

And another distraction would come with the outbreak of war in 1939.

War Service

Du Toit was an unequivocal supporter of the Allied effort during the Second World War. Initially, this support was pecuniary with donations made to the Johannesburg Fund.[4] But, as the war progressed, Du Toit was pulled out of retirement to play a more active role. In 1943, at the age of 65, he continued his work in the field in aid of the war effort in the northern part of the Union in the area of what he termed "mineral production" declaring with some sense of irony that his retirement—far from providing the freedom to undertake research as he had hoped—had inflicted him with more duties than he had undertaken when still in paid employ.[5] Du Toit's work during the war involved asbestos which he believed to be of significance to the war effort, "[it was] a substance the enemy would dearly like to get more of", and necessitated intermittent bouts of fieldwork—no small undertaking for a man of advanced age.[6] He had also focused on asbestos as a significant part of his research after his retirement from De Beers in 1941, attempting to understand the conditions contributing to the formation of asbestos deposits that culminated in the publication of a paper in the *Transactions of the Geological Society of South Africa* in 1946.[7]

The Second World War allowed for an understanding of Du Toit's political sympathies and sense of identity as a white South African of

[3] UCT-JL: Alex L. Du Toit Papers—BC 722: A4—Personal Documents: Handwritten CV/Brief Biography.

[4] "£500 City Gift to War Fund," in *The Rand Daily Mail*, Saturday, 3 August 1940 and "Another £600 for Raid Fund," in *The Rand Daily Mail*, Thursday, 17 October 1940.

[5] UCT-JL: Alex L. Du Toit Papers—BC 722: B1—General Correspondence (1930s–1940s). Letter by Du Toit to RW Newman, 4th May 1943.

[6] UCT-JL: Alex L. Du Toit Papers—BC 722: B1—General Correspondence (1930s–1940s). B3.2 Correspondence: D-H—Letter by Du Toit to Robert Hamilton of the U.S. Armed Forces, 1 March 1944.

[7] Haughton, "Alexander Logie Du Toit, 1878–1948," p. 390.

Afrikaner and Huguenot descent. What was remarkable about Du Toit's writings, particularly at the outset of his career, was the lack of political commentary—despite witnessing the tumult that accompanied the creation of a nation out of disparate, competing and often hostile elements. His diary for 31st May, 1910, for instance—the day that the Union of South Africa came into being—focuses on the mundanity of fieldwork preliminaries to the exclusion of all else, involving the habitual commentary on the weather ("fine day but cold"), the activities of tea-drinking, letter-writing and preparing the wagon.[8] This absence is just as marked in 1915 as Du Toit carried out his war service in the Union Defence Force locating groundwater supplies in South West Africa with only the occasional mention of the rank of acquaintances hinting at the conflict and his official role.[9] Yet, just as these events were moments of great stress for the fledgling nation, the same was true a generation later with the outbreak of the Second World War and, here, Du Toit was less reticent.

The Second World War highlighted the divisions between both English- and Afrikaans-speaking white South Africans evident in Jan Smuts's United Party and the rising tide of Afrikaner Nationalism represented by the National Party—which would ultimately dominate in 1948.

Du Toit's participation in the war—albeit in a limited capacity not least due to his age—was a clear statement of intent. From the beginning of September, he made brief reference to the events leading to South African involvement—Germany's invasion of Poland, the subsequent declaration of war by the Allies and, finally, the formation of a new cabinet under the leadership of Smuts.[10] As the war progressed, his indictment of the subversive activities of the right-wing extremists was unambiguous: "Nazism I regard as the syphilis among nations; indeed it is hard to obtain terms sufficiently harsh & brutal for it. Out here we have a large proportion who seem to have plumped for Nazi rule & all that such will mean. For such leaders there should be life imprisonment or the hangman's rope".[11] Of interest to him as well, was a statement issued by a number of academics at Stellenbosch University—the alma mater of Jan Smuts, JBM Hertzog

[8] UCT-JL: Alex L. Du Toit Papers—BC 722: A1—Diaries, 1910—31 May 1910.

[9] UCT-JL: Alex L. Du Toit Papers—BC 722: A1—Diaries, 1915—10 April 1915.

[10] UCT-JL: Alex L. Du Toit Papers—BC 722: A1—Diaries, 1915—3 and 5 September 1939.

[11] UCT-JL: Alex L. Du Toit Papers—BC 722: B1—General Correspondence (1930s–1940s): Letter by Du Toit to Mr. La Tendresse, 19 November 1940.

and DF Malan—rejecting right-wing extremism as antithetical to Afrikaner values.[12]

As 1941 drew to a close, brief mention was made in *The Star* of a commemorative plaque to be laid by the Historical Monuments Commission commemorating the site where a number of Du Toit's Huguenot ancestors were buried from the late seventeenth century.[13] Contained amongst Du Toit's papers, it indicated a sense of pride in his ancestral origins and an adoption of an identity that was both Afrikaner and South African and just as inextricably linked to the liberalism that exemplified the Allied war effort. It represented a harmonisation between Du Toit's political and intellectual life.

Du Toit's war service, then, was based on duty and patriotism despite the unwelcome distraction it presented from research. Yet, even as Du Toit occasionally displayed certain cantankerousness, the demands upon his time were based on an appreciation of his expertise and dedication to the field of geology—which would also have more concrete and less onerous implications.

Awards and Accolades

Du Toit was awarded the Murchison Medal in 1933,[14] by the Geological Society, an award "given to someone who has made a significant contribution to the science by means of a substantial body of research".[15] The same year would mark an acknowledgment of his work in geology by his peers in South Africa. In his biography of Du Toit, Gevers believed that it was only during the later years of his career that Du Toit received sufficient recognition in South Africa—in contrast to his many international accolades. One of these was the David Draper Memorial Medal which was given to Du Toit in 1933.[16] In a sense this signified a passing of the

[12] UCT-JL: Alex L. Du Toit Papers—BC 722: K—Miscellaneous: "Stellenbosch speaks up for democracy; Totalitarian challenge to Afrikaner Heritage; Manifesto by University Professors," [Publication Unknown] 15 April 1941.

[13] UCT-JL: Alex L. Du Toit Papers—BC 722: A4—Personal Documents: Excerpt taken from *The Star*, 10 December 1941.

[14] Haughton, "Alexander Logie Du Toit, 1878–1948," p. 391.

[15] http://www.geolsoc.org.uk/About/Awards-Grants-and-Bursaries/Society-Awards/Murchison-Medal. Accessed 1 February 2016.

[16] Gevers, "The Life and Work of Dr. Alex. L. Du Toit," p. 29.

torch—David Draper's career had ended just as Du Toit's was on the ascent.

Fittingly, then, on receiving this award Du Toit was aware of the challenges that each succeeding generation would bring to the establishment: "Looking back, one sees that in geology, as in every other science, the discoveries of tomorrow inevitably modify or else supersede the conclusions of yesterday, and the most that one can pray for is that some few of one's youthful expressions of geological opinion may manage to elude the searching test of time. Only in such a way, however, can geological knowledge progress!"[17] For him, there were no sacred cows and scientific progress could only come through challenging existing views that had become dogmatic.

In 1941, Smuts was approached by Sir Thomas Holland, in his youth the Assistant Superintendent of the Geological Survey of India, elected president of the British Association of the Advancement of Science during its South African visit in 1929 and current Principal and Vice Chancellor of Edinburgh University.[18] Despite his criticism of *Our Wandering Continents* as being a challenging read, Holland had nominated Du Toit for membership to the Royal Society. In his letter to Smuts, Holland pointed out that numerous geologists had been nominated and there was a tendency to favour more youthful scientists. He therefore suggested that Smuts's endorsement of Du Toit would help the latter's chances of admission into the society.[19] Smuts's reaction was unequivocal. In a letter to the Secretary of the Royal Society, Smuts highlighted Du Toit's contribution as a geologist dealing with the pitfalls of working away from the metropole who had nevertheless managed to make a valuable contribution to the field of geology as a whole:

> I consider Dr A.L. Du Toit eminently qualified for membership ... I believe it is sound policy from wider points of view to give recognition to scientific workers of high merit in the Dominions, who often work under great discouragement and disabilities in the interests of Science. Dr A.L. Du Toit is

[17] UCT-JL: Alex L. Du Toit Papers—BC 722: A4—Personal Documents: "Proceedings of the Geological Society of South Africa," p. li.

[18] L.L. Fermor, "Thomas Henry Holland, 1868–1947," in *Obituary Notices of Fellows of the Royal Society*, Vol. 6, No. 17 (Nov. 1948), pp. 83, 89, 91.

[19] UCT-JL: Smuts Papers Vol. 65, No. 41: Letter by T H Holland to Smuts, 18 August 1941.

> not only an outstanding South African scientist, but his original contributions to Geology are also outstanding.[20]

On the 18th of March 1943, "Alexander Logie Du Toit, lately consulting geologist to the De Beers Consolidated Mines" was named a fellow of the Royal Society in London. Those elected reflected the diverse areas of scientific research, ranging from bacteriology to astronomy. Another geologist elected was Guy Ellcock Pilgrim who had held the position of Superintendent of the Geological Survey of India. What was particularly noteworthy about the list of "New Fellows" was that they reflected the trend in British research that included the Empire. Honoured along with Du Toit was Sir Shanti Swarupa Bhatnagar of India and Wilder Penfield and John Lighton of Canada.[21] Simultaneously, however, Du Toit's response to the Royal Society reflected his position as a representative of South African scientific endeavour, echoing Smuts's own sentiments:

> I should be greatly obliged if you would convey to the President & Council my deep appreciation of the immense honour done not only to me personally but to scientists in the Union of South African generally in my election to the Royal Society, a distinction long aspired though hardly anticipated. The warmth of the congratulations from my colleagues in this country forms an eloquent testimonial of the high regard in which this award is held among scientists.[22]

Days later Du Toit received an honorary D. Sc. from the University of the Witwatersrand:

> In his speech the Principal jokingly said that the University of the Witwatersrand was as much honoured by the Royal Society as Dr Du Doit, for hardly had that august body learnt that the University was about to bestow an Honorary D.Sc degree on Du Toit, than they promptly forestalled the University by electing him a Fellow. The University at least now knew that they had not bestowed their own degree on an undeserving fellow.[23]

[20] UCT-JL: Smuts Papers Vol. 66, No. 177: Letter by Smuts to the Secretary, Royal Society, 11 October 1941.

[21] "The Royal Society: New Fellows Elected," in *The Times,* March 19, 1943, Issue 49498, p. 2. London, Times Newspapers Limited. Gale Document Number CS34815091.

[22] UCT-JL: Alex L. Du Toit Papers—BC 722: B3 Correspondence: B3.3 Correspondence (c.1937–1947): Letter by LC King (Natal University College) to Du Toit, 2 May 1944.

[23] Gevers, "The Life and Work of Dr. Alex. L. Du Toit," p. 30.

The principal's humorous interjection displayed the complex relationship between former colony and metropole, an apt symbol of Du Toit's work on continental drift: South Africa had pre-empted the Royal Society by deciding to honour Du Toit but their action was itself validated by the actions of the Royal Society.

Du Toit had received honorary doctorates from various South African institutions including the University of Pretoria and the University of South Africa but the ceremony at the University of the Witwatersrand placed him in distinguished company. Isie Smuts, wife of Jan Smuts, was also being honoured with an honorary doctorate in law as a result of her efforts during the Second World War. Du Toit's degree was presented by T.L. Gevers of Wits' Department of Geology—who would go on to write a brief biography of Du Toit after the latter's death. The opening address was given by J.H. Hofmeyr, chancellor of the university who, within the context of the nation-building efforts in the midst of war, highlighted the common destiny of the various groups in the country with an emphasis on "progress" for all. Particularly emphasised was the valuable European cultural input from both Afrikaans and English-speaking white South Africans that would comprise the "national contribution to the common stock of humanity". The theme of settler intellectual dominance was reiterated in Smuts's address who employed the metaphor of the Great Trek—settler expansion (and domination):

> We have left the old securities, the old secure places, and are once more on trek on a great search for the future. The old spirit of the march has once more seized on mankind, and the movement to some far-off goal has once more begun ... We honour our pioneers, our Voortrekkers. No less will the future honour these Voortrekkers of a later generation, whose footprints are over all Africa. They have staked their and our claim on the future, and earned their title to us whatever good may be in store for us.[24]

Legacy

When I first set out to write this book, the figure of Alex Du Toit was almost secondary to the story that I wanted to tell—that of the history of plate tectonics with a particular focus on South Africa and the ideological

[24] UCT-JL: Alex L. Du Toit Papers—BC 722: A4—Personal Documents: Article appearing in *The Star*, 20 March 1943.

intellectual milieu of the segregation era. I deliberately put off reading the only detailed biography of Du Toit written by T.W. Gevers in honour of the inaugural Alexander L. Du Toit Memorial Lecture as I wished to understand Du Toit largely through his work as a geologist. At the same time, however, I was hardly unaware of one of the most common portraits of Du Toit—long, thin face, high forehead and a certain quirk of the lips that indicated good humour (Fig. 11.1).

This was borne out by the various tributes paid to him by colleagues and friends who depict a modest man with ascetic tendencies who nevertheless had a passion for life and an insatiable curiosity. He loved art and music—playing the oboe—and even took up motorcycle racing in his later years. As an academic and teacher, he was fair in his criticism and treated both fellow intellectual and novice student alike with the same respect.[25]

Yet Du Toit as a man was almost outdone by Du Toit as geologist. In a period of specialisation, his geological expertise was breath taking and matched only by his passion for the field.

Alex Du Toit died on the 25th of February 1948 at the age of 69 following a brief illness. At the time of his passing he was about to become president of the Royal Society of South Africa, he was the chairman of the Council of the South African Archaeological Committee and had been involved in attempting to create an umbrella organisation for scientific societies in the Cape Province. He had also edited a commemorative volume on Robert Broom for the Royal Society of South Africa. At the end, then, his life was as busy and his interests as diverse as they had been throughout.[26]

It was only a few months after Du Toit's death that South African scientists and intellectuals took concrete steps to honour his work. A fund was started by the Royal Society of South Africa to assist in funding the areas of research in which Du Toit was engaged, including the Athenaeum project and to be used as an award given each year "for the publication of an outstanding scientific paper in geology, irrigation or archaeological chronology".[27]

[25] S.H. Haughton, "Alexander Logie Du Toit, 1878–1948," in *Obituary Notices of Fellows of the Royal Society*," Vol. 6, No. 18 (November 1949), pp. 391–392.

[26] A.G. "Obituary: Alexander Logie Du Toit, F.R.S.," in *the South African Archaeological Bulletin*, Vol. 3, No. 9 (March 1948), pp. 14–15.

[27] "Fund in Memory of SA Scientist," in *Rand Daily Mail*, Saturday, 24 July 1948. http://phw02.newbank.com/cache/arhb/fullsize/pl_010272016_0557_01373_303.pdf. Accessed 27 October 2016.

Fig. 11.1 Alexander Logie Du Toit (University of Cape Town Libraries, mss_bc722_ql)

A little more than a year later in October 1949, the inaugural "Alex L. Du Toit Memorial Lecture" was given by the University of Witwatersrand's Professor of Geology, Dr T.W. Gevers. It was appropriately entitled "The Life and Work of Alex L. Du Toit" and was a significant biography of the respected geologist. The lecture marked the first in a series that would take place every two years, based on the many subfields of geology in which Du Toit was involved and involving both national and international speakers.[28]

Yet, Du Toit's career and, arguably, his greatest contribution to geology—continental drift—was marked by controversy during his lifetime. While he had his South African supporters, such as Lester King and drift theory was supported by Jan Smuts, his South African peers were more circumspect, as evident in the Gevers' ambivalence. Du Toit's strength as a geologist—his fieldwork, mapping and wide-ranging knowledge were both recognised and respected. This did not extend to what would become his defining work—continental drift. The same applied to the international reception of continental drift where validation of his work was balanced by hostile criticism. It would only be in the decades after his death that Du Toit's work would be truly appreciated as it was incorporated into plate tectonics.

An "Alex L. Du Toit Memorial Lecture" given in 1961 deserves special mention. Henno Martin was a German geologist. He, along with fellow geologist Hermann Korn were graduates of the University of Göttingen and had moved to South West Africa (present-day Namibia) in an attempt to put some distance between themselves and Nazi Germany. Here they were involved in locating water resources—a very necessary project in an arid region—and acquired a good understanding of the geology of the area. At this point South West Africa was a mandate of South Africa and, in a cruel twist of fate, upon the outbreak of war, both men—fearing that they would be interned by the South African state—fled into the desert. Their necessary exile in hostile terrain also provided them with the opportunity for geological work and they were able to accumulate evidence for continental drift based on the work of Wegener and Du Toit, with a particular focus on glaciation. Korn was subsequently killed soon after the end of the war but Martin continued work in Brazil, focusing on correlating sequences between South America and southern Africa, as Du Toit had done. It was fitting, then, that Martin's lecture "The hypothesis of continental drift in the light of recent advances of geological knowledge in Brazil and South West Africa" was based on his (and Korn's) work in Namibia as well as South America. It was an acknowledgement of Du Toit's legacy just as continental drift was about to become an overarching

[28] "SA Scientist to be Honoured by Series of Lectures," in *Rand Daily Mail*, Thursday, 20 October 1949. http://phw02.newbank.com/cache/arhb/fullsize/pl_010272016_0550_15136_454.pdf. Accessed 27 October 2016.

explanatory framework for geological processes in the form of plate tectonics.[29]

As evidence mounted for plate tectonics in the 1960s, Alex Du Toit assumed new significance. While greater focus has often been accorded to Alfred Wegener, Du Toit is nevertheless acknowledged as a key figure who provided stratigraphic, glacial and fossil evidence on which later geologists would draw. Frankel's comprehensive four-volume account of the history of continental drift and plate tectonics pays due consideration to Du Toit. He is also addressed briefly in Oreskes' edited collection, *Plate Tectonics: An Insider's History of the Modern Theory of the Earth*. Geologists at the forefront of plate tectonics were familiar with the work of both Wegener and Du Toit. And the ultimate acceptance of plate tectonics would cement Du Toit's reputation as a prescient figure.

Du Toit's contribution to geology both in his own country and internationally was even given extra-terrestrial recognition in 1973 when a crater on Mars was named after him.[30] In October 1991, the South African Postal Service issued a series of stamps entitled "South African Scientists" which featured four scientists. Alongside Basil Schonland and Robert Broom was Alex Du Toit, acknowledged for his geological mapping of the Karoo, his stint with De Beers and for his contribution to continental drift with *Our Wandering Continents*.[31]

The 35th International Geological Congress (IGC) was held in Cape Town, South Africa, in late August and early September 2016. Sandwiched between the dramatic backdrop of Table Mountain and the Atlantic Ocean, its location was particularly apt and seemed to appeal to the geologists too with thousands of delegates attending and hundreds of papers presented. I was, of course, drawn to that which was most familiar to a graduate of the human sciences—geohistory, geoethics and geoeducation. In the present, plate tectonics is understood as an all-encompassing process that underlies much of geological research and looming large in discussions of the history of plate tectonics were the figures of Alex Du Toit, Alfred Wegener, Juan Keidel and Eduard Suess.

[29] Ted Nield, *Supercontinent: Ten Billion Years in the Life of Our Planet* (London, Granta Books, 2008), pp. 126–128.

[30] https://ipfs.io/ipfs/QmXoypizjW3WknFiJnKLwHCnL72vedxjQkDDP1mXWo6uco/wiki/Alexander_du_Toit.html. Accessed 27 July 2018.

[31] http://www.paleophilatelie.eu/description/stamps/south_africa_1991.html. Accessed 11 June 2020.

While Du Toit had achieved international recognition (and, at times, criticism) for his work on continental drift, he had also made an indelible contribution to the discipline of geology in South Africa. From his early career involving strenuous fieldwork in unforgiving conditions, he had acquired an expertise in the field that transcended the boundaries of specialisation and this expertise would be lent to areas ranging from mentorship to the extraction of minerals, hydrogeology and war. The controversial nature of drift theory did little to detract from his peers' recognition of him as one of the foremost geologists of the first half of the twentieth century.

On the 22nd of May, 1948, three months after Alex Du Toit's death, the following anecdote appeared in Leonard Bright's column, "Men, Women, Events: A Daily Causerie":

> More than two years ago Professor Du Toit invited me to tea one Sunday morning at his home in Pinelands in the Cape. After tea I said, "You have some lovely paintings here. That's a grand Pierneef." He said, "I like it, too. You see that outcrop on the right? That's decomposed dollarite [*sic*], just as happens in those parts." I pointed to another and said, "A good Volschenk." "You're right," he told me. "That cliff, and the configuration of those rocks, show the Bokkeveld series."

That is perhaps the best epitaph for Alexander Logie Du Toit—a man who saw the masterpiece of geology.[32]

[32] Leonard Bright, "Men, Women, Events: A Daily Causerie," in ***Rand Daily Mail***, Saturday, 22 May 1948. http://phw02.newbank.com/cache/arhb/fullsize/pl_010272016_0606_32763_126.pdf. Accessed 27 October 2016.

CHAPTER 12

Pale Blue Dot: Conclusions

As part of the opening remarks of his public lecture given in July 1921, Alex Du Toit stated:

> The principle of splendid isolation appeals universally to mankind, and therefore it may be a little disconcerting at first to learn that in the dim past the African continent, geographically and biologically, was linked at times to the several other land-masses of the globe.[1]

Du Toit was clearly by no means unaware of the ideological importance placed on the notion of space but his work in what would become plate tectonics and that of the discipline of geology itself has a complex relationship with space which has shaped the ways in which the knowledge of Earth processes was acquired.

The title is taken from Carl Sagan, *Pale Blue Dot: A Vision of the Human Future in Space*. (New York: Random House, 1994). In his book Sagan presents a picture of the fragility and isolation of the earth when viewed from space, with a need for human unity.

[1] Alex L. Du Toit "Land Connections between the Other Continents and South Africa in the Past", Public Lecture given on 15 July, 1921, p. 120. Published in *South African Journal of Science*, 18, 120–140.

S. Chetty, *'Africa Forms the Key'*, Cambridge Imperial and Post-Colonial Studies, https://doi.org/10.1007/978-3-030-52711-2_12

The Dark Continent?

In a manner that pre-empted knowledge of Pangaea, medieval European perceived little geographical difference between the African, European and Asian continents. Early cartographers depicted Africa, Europe and Asia as a monolithic continent, broken only by the Mediterranean and Red Seas that served to distinguish Africa's northern and eastern boundaries. This comprised *Ecumene* or "the known world" beyond which lay a watery equator separating the northern hemisphere from the mysterious *Antipodes*. While knowledge of Africa was greater in the Arab world due to the incursions into East Africa,[2] it was European voyages in the late fifteenth century that would comprehensively render the outlines of the African continent and mark its separation from Europe, both geographically and psychologically.

Scientific laws were unchangeable—they were constant across time and space. Science, then, was not perceived as being contextualised by space yet the reception of science, the ways in which it was applied, the determinations over what was appropriate to research and what was not were very much based on space—both physical and ideological. The voyages of discovery brought the terra nova of the Americas into European consciousness, the notion of "darkest Africa" in the nineteenth century subjected the continent to the "civilising" influences of science and progress, the Orient became the site of European cultural, scientific and military influence. And, accompanying all of this was exploration, surveying, mapping, the will to know, to understand, to name and, ultimately, to control.[3]

While the new scientific methodology prioritised disinterested direct observation and a scepticism towards all received knowledge, as a greater portion of the globe fell under the European gaze, it was no longer possible for scientists to have first-hand experience of its more remote parts. Knowledge at the metropole had to be constructed from accounts sent by the vanguard of empire—explorers, travellers, missionaries and, eventually, settlers. They were trained in the scientific methods enabling them to be good, impartial observers and provided with the instrumentation to aid them in this endeavour. Also used were maps which were particularly associated with impartial truth yet the rendering of the three-dimensional

[2] Michael F. Robinson, *The Lost White Tribe: Explorers, Scientists, and the Theory that Changed a Continent.* (New York, Oxford University Press, 2016) pp. 25–26, 27.

[3] David N. Livingstone, *Putting Science in its Place: Geographies of Scientific Knowledge.* (Chicago and London, The University of Chicago Press, 2003) pp. 8–10.

Earth into a two-dimensional scaled space that simplifies and excludes is itself a process of interpretation.[4] It was in the space between reality and representation that Wegener—and Du Toit's—interpretation of the fit between South America and the west coast of Africa lay.

As the British Empire expanded in the nineteenth century, it did so under the guise of "improvement"—both moral and scientific. The rhetoric of improvement provided the means by which a common imperial identity could be created among the various colonies and the British core. It allowed the accrual of economic, political and strategic benefit to Britain while possibly improving the lives of those who fell under British control.[5] Improvement also became the means by which imperial domination was vindicated, particularly when compared to those European powers who failed to deliver on the "civilising mission". In South Africa, for instance, criticism was aimed at the previous Dutch colonial administration which had failed to introduce "improvements…into the colony" due to their "want of energy, and…natural indolence", in contrast to British enterprise and solicitude for those it governed.[6]

Sir Joseph Banks, president of the Royal Society, believed that the principles of improvement—the benefits of free trade, that would primarily benefit Britain, and the "civilising mission"—could be implemented in the colonies without the attendant political aspirations of self-government and nationalism: "You need not be afraid of the colonies, if you do not furnish them with the elements of independence as you did with North America".[7] But Britain was not as adroit at controlling these boundaries as Banks believed possible. Independence would come and, with it, a challenge to the metropole as the centre of scientific knowledge. Alex Du Toit would be part of a later generation who would benefit from improvement—born soon after the Cape Colony became a self-governing territory, educated in the metropole and in his thirties when the Union of South Africa was formed, he exemplified the unforeseen consequences of imperialism, science and the Enlightenment.

Du Toit harnessed the European-derived discipline of geology to the definition of a new nation and then proceeded to assert the primacy of the

[4] Livingstone, *Putting Science in its Place*, pp. 140–153, 154–155.

[5] John Gascoigne, *Science in the Service of Empire: Joseph Banks, the British State and the Uses of Science in the Age of Revolution*. (Cambridge, Cambridge University Press, 2010) p. 169.

[6] Gascoigne, *Science in the Service of Empire*, p. 170.

[7] Gascoigne, *Science in the Service of Empire*, p. 178.

nation—whether it was as the originator of life or the centre of Gondwana. In addition, through the survey of the landscape for the natural resources, necessary both for the development of the nation and for the imperial economic network (such as Du Toit's involvement early on in determining underground water and coal supplies as well as his work on diamond-bearing deposits and, later, gold), it became possible to understand the geology of the landscape, contextualising it both nationally and globally. Canadian geologist, Richard Dawson would relate geology to the colonial project: "since 'geological conditions…determine beforehand the resources and population' of any given land, it was geologists who could best inform settlement and development policy".[8]

As members of a settler population—and thus the representatives of imperialism and colonialism—on the southernmost tip of Africa, men like Du Toit were in a marginal position. This was compounded by their status as citizens of an ostensible democracy that excluded a vast majority of the country's inhabitants from equal political participation. Their perspective of Africa was thus contextualised by nationalism, imperialism and paternalism. Du Toit's assertion in *Our Wandering Continents* that Africa was "the key" was a product of nationalism, a declaration for the importance of the African continent but an importance that had been determined by science that had its origins inescapably in the European Enlightenment. Even as he emphasised the importance of the continent in the history of the Earth (and of humankind), it was predicated upon settler intellectual—and political—leadership. The reaction by Charles Schuchert prioritised another view of Africa. A first generation American of German descent, Schuchert's amassed a formidable collection of fossils that brought him to the attention of palaeontologist, James Hall, who subsequently hired him as an assistant. From these fairly humble beginnings, Schuchert was eventually made Professor of Paleontology at Yale University and Curator of the Geological Collections in the Peabody Museum in 1904. By the time of his writing to Du Toit he had long since retired and assumed the distinguished role of Professor Emeritus, allowing him the freedom to engage in research. Through his expertise in fossils and ability to synthesise vast quantities of information, Schuchert was able to reconstruct past environments and was renowned for his creation of palaeogeographic maps that

[8] Suzanne Zeller, "The Colonial World as Geological Metaphor: Strata[gems] of Empire in Victorian Canada" in *Osiris*, 2nd Series, Vol 15, "Nature and Empire: Science and the Colonial Enterprise" (2000), p. 105.

emphasised the advance and retreat of the seas over the North American continent. Related to this was orogeny and Schuchert was a firm proponent of the theories regarding mountain-building advocated by his early mentor, James Hall.[9] As such, he was disinclined to accept the views put forth in *Our Wandering Continents.*

After expressing his surprise at having been sent a copy of the book, Schuchert regretted that he was unable to return the kindness due to his firm disbelief that continents had "floated" over vast distances. Admiring Du Toit for the frank expression of his views—despite those views standing on a "sliding base"—the palaeontologist nevertheless viewed Du Toit's interpretation as confirming the marginalisation and isolation of Africa rather than its centrality: "To you, 'Africa forms the Key', and your unlocking of the drift theory permits us to see your view of how the continents of Gondwana have flown away from Darkest Africa". Clearly demonstrating a mastery of the double entendre, Schuchert's language is significant in its use of the phrase "Darkest Africa", signifying the dismissal and negative stereotype of the previous century.[10]

Geologists were hostages to their geology. In the case of American geologists, the geology of the continental United States as well as their emphasis on a heavily inductive scientific methodology made many less kindly disposed towards continental drift theory—in contrast to their European counterparts such as the Dutch whose colonial experience had exposed them to vastly different geological landscapes. Du Toit was no less a product of geology and ideology yet his view of continental drift was shaped by his expertise in South African geology, his travels abroad to the regions that once comprised Gondwana, his correspondence with "southern" scientists who held similar views and, finally, an ambitious vision unlimited by national boundaries leading to the construction of a whole out of disparate parts. Simultaneously, however, he was an integral part of an ideological climate that asserted South African intellectual leadership and dominance on the African continent. It was this merging of the global and the national, the scientific and the ideological that would enable Alex Du Toit to claim, "Africa forms the key".

[9] Adolph Knopf, "Charles Schuchert: 1858–1942" (Washington D.C., National Academy of Sciences, 1952) http://www.nasonline.org/publications/biographical-memoirs/memoir-pdfs/schuchert-charles.pdf, Accessed 12 September 2017, pp. 364, 366, 368–369, 372–373.

[10] UCT-JL: Alex L. Du Toit Papers—BC 722: B1—General Correspondence (1930s–1940s)—Letter by Charles Schuchert to Du Toit, 20 December 1937.

Postscript: Plate Tectonics and the Vindication of Drift Theory

At the time of Alex Du Toit's death in 1948, continental drift was far from accepted. In his brief biography of Du Toit, published a year later, T.W. Gevers was himself wary of the concept. It would take another two decades and a preponderance of evidence before the theory would enter mainstream geology as plate tectonics.

The Mechanics of Tectonics

Plate tectonics theory had its origins in conflict. A fundamental criticism aimed at continental drift was that the evidence cited for (and against) it was largely circumstantial. The interior of the Earth and its workings were confined to the realm of conjecture, the repository of ideology and mythology. It was, ironically, conflict on a global scale that provided scientific evidence for plate movement and the once—and future—interconnected nature of continents. The history of plate tectonics, however, is not necessarily a linear narrative but instead consists of putting together the various pieces of a jigsaw puzzle until a whole picture is formed—an analogy that Alex Du Toit, with his reconstruction of Gondwana, would have appreciated.

The depredations of German U-boats on Allied shipping during the Second World War prompted the United States Navy to investigate the ocean depths. This research took two forms—the first was in magnetism which could serve as a means of tracking submarines and the other was in

S. Chetty, *'Africa Forms the Key'*, Cambridge Imperial and Post-Colonial Studies, https://doi.org/10.1007/978-3-030-52711-2

ocean topography in order to evade them. Harry Hess was appointed captain of the USS *Cape Johnson* and tasked with using echo-sounding to map the sea floor of the Pacific.[1] Soon after the Second World War, the new technology was being employed to better understand the ocean floor. A moving remnant among Du Toit's papers is a newspaper clipping dated just months before his death. It detailed the exploration of the Mid-Atlantic Ridge undertaken by the National Geographic Society. The expedition had taken two months and was led by Professor Maurice Ewing of Columbia University aboard the *Atlantic*, a research ship. The geologists had identified a particular portion of the ridge almost two thousand kilometres from Bermuda where rock samples were taken in order to be analysed later as a means of determining their origins and correlation with other parts of the world. The expedition had also realised the dramatic landscape that composed the ridge with "sharp peaks and valleys in the submarine range, some peaks being about two miles high rising abruptly from the flat ocean floor".[2] This was yet another piece of the puzzle and it would be Harry Hess who would paint a compelling portrait of the ocean depths and its significance for plate tectonics.

In 1931, Hess had worked on measuring gravity in the ocean. This had formed part of ongoing research since the 1920s where scientists investigated whether the principle of isostasy could be applied to the sea, just as it had to the land. The resulting measurements suggested to Hess that the ocean floor was active, concluding that there were deformations in the ocean crust that could be discerned as trenches.[3] He continued his research on board a naval submarine and subsequently published a paper on the island arcs that were usually associated with trenches[4]—what are known today to be the visible signs of subduction zones.

Hess eventually captained the USS *Cape Johnson* which was assigned to the Pacific where he was involved in heavy fighting as American forces staged landings on islands such as Iwo Jima. Even as the military

[1] Naomi Oreskes, "From Continental Drift to Plate Tectonics" in *Plate Tectonics: An Insider's History of the Modern Theory of the Earth*. Naomi Oreskes, (ed). (Boulder, Westview Press, 2003) pp. 16–17.

[2] UCT-JL: Alex L. Du Toit Papers—BC 722: C1—Research Papers Relating to his Book "Our Wandering Continents": C1.4—Geodetic: "Probing Secrets of Great Mountain Range Under the Sea" [Newspaper title unknown], 19 September 1947.

[3] Oreskes, "From Continental Drift to Plate Tectonics", pp. 13–14.

[4] Harold L. James, "Harry Hammond Hess, 1906–1969: A Biographical Memoir". (Washington D.C. National Academy of Sciences, 1973) p. 112.

campaigns in the Pacific played out, the *Cape Johnson* was also involved in scientific research. Echo-sounding allowed Hess to identify—and name—guyots, underwater mountains rising from the sea floor with flat summits. Hess' particular interest was in peridotite—and he even travelled to South Africa in 1969 to study the lava flows of the Barberton Mountains that appeared to be of largely peridotite origin. Hess believed that these rocks which originated in the Earth's mantle were important to understanding the nature of the Earth's crust and it was this fascination with peridotite as well as his geological findings during the Second World War that were to play an important role in understanding plate tectonics.[5]

Across the Atlantic, Teddy Bullard and P.M.S. Blackett would focus on magnetism, Bullard's research in particular contextualised by the Second World War where he focused on magnetism and its potential applications in ameliorating the threat to ships posed by magnetic mines.[6] He hypothesised that the Earth's magnetic field was a result of the electrical dynamo generated by currents within the Earth's core (which he believed to be liquid) that created the magnetic field. These shifting currents suggested a "transient" magnetic field the strength of which would fluctuate over time. Bullard had also focused his attention on the mid-ocean ridges, showing that the temperature here was at its height in relation to the rest of the ocean floor. This suggested "rising convection currents".[7]

Blackett, along with an associate, S.K. Runcorn, proposed that, by using Bullard's hypothesis, they would be able to trace a record of the changing magnetic field in rocks. As rocks cooled, the magnetic elements within them are magnetised in the direction of the magnetic north pole. Blackett, along with Runcorn and his assistant, Edward Irving, found that the direction of magnetism varied among rocks of different age. On the one hand, this could suggest that the poles themselves had wandered—a theory evident in the early twentieth century. On the other hand, if continents containing rocks of the same age showed that the direction of magnetism within these rocks varied across continents, this suggested that the continents themselves had moved and were in different positions in relation to the poles when the rocks were first formed. In 1956, fieldwork in

[5] James, "Harry Hammond Hess", pp. 112, 114.

[6] Oreskes, "From Continental Drift to Plate Tectonics", p. 18.

[7] Oreskes, "From Continental Drift to Plate Tectonics", pp. 18–19, 20.

India proved that this was in fact the case and a strong case could be made for continental drift.[8]

Based on the conclusions drawn by the history of magnetism preserved in rock—or palaeomagnetism—and their repercussions for continental drift, Harry Hess wrote a ground-breaking paper in 1962 that proposed that rising convection currents provided the mechanism for continental drift, renting the crust asunder at the mid-ocean ridges and drawing the crust down at the ocean trenches.[9] In a concession to the still-vociferous drift detractors, Hess described his paper "History of Ocean Basins" as an "essay in geopoetry".[10] This "geopoetic" exercise would become the most cited paper in "solid-earth geophysics" in the late 1960s.[11] For the detractors, however, more proof was needed and the puzzle pieces would start coming in until the evidence was largely irrefutable.

Again, magnetism was key. The Scripps Institution of Oceanography developed a magnetometer that would allow them to record the magnetism of the ocean floor. In 1955, the *Pioneer*, a ship attached to the US Coast and Geodetic Survey was involved in attempting to map the sea floor off the western United States, a project that was estimated to take two years. This was the perfect opportunity for Scripps to use its magnetometer and members of the Scripps Institution joined the *Pioneer* in August 1955, for the first of 12 voyages that would end in October a year later. Once the data had been assembled and the map completed in 1957, it revealed alternating bands or "stripes" of positive and negative magnetism. The reasons for this, however, remained open to conjecture for the next five years until Frederick Vine and Drum Matthews provided a compelling, yet simple, explanation.[12]

Frederick Vine was a teenager in West London in 1955 and, at the tender age of 15, decided that there needed to be a suitable account of continental drift, "It seemed to me that one could hardly conduct any

[8] Oreskes, "From Continental Drift to Plate Tectonics", pp. 19–20.

[9] Oreskes, "From Continental Drift to Plate Tectonics", p. 20.

[10] Oreskes, "From Continental Drift to Plate Tectonics", p. 21 and James, "Harry Hammond Hess", p. 125.

[11] James, "Harry Hammond Hess", p. 115.

[12] Ron Mason, "Stripes on the Sea Floor" in *Plate Tectonics: An Insider's History of the Modern Theory of the Earth*. Naomi Oreskes, (ed). (Boulder, Westview Press, 2003) pp. 32–35, 36, 40.

meaningful history of the Earth until one had resolved this issue".[13] Vine was awarded a state scholarship to study at St John's College, Cambridge where he studied geology and developed a particular interest in geophysics—specifically as it related to continental drift. He saw a growing tolerance of continental drift in the United Kingdom (and the British Commonwealth) that was in stark contrast to the United States.[14]

At Cambridge, Vine became Drummond H. Matthews' research assistant. At the time Matthews was part of the International Indian Ocean Expedition and acquired a wealth of information related to the magnetism of the Indian Ocean floor and was involved in a detailed survey of the Carlsberg Ridge located in the northwest.[15] It was his role in interpreting Matthews' findings that set Vine on the path towards the "Vine-Matthews hypothesis".[16] Vine was just 24 years old when "Magnetic Anomalies over Oceanic Ridges" was published in *Nature* in September 1963. The paper made two significant points that were subject to contestation: underwater ridges were the site of seafloor spreading and the magnetic field that surrounded the Earth reversed periodically. These reversals were captured in the palaeomagnetic rock record and were indicated by the alternating stripes of positive and negative magnetism first revealed by the *Pioneer*. As magma reaches the surface during volcanic eruptions, magnetic/metallic elements within them align with the magnetic north pole and this alignment is preserved as the lava cools and solidifies, thus indicating the position of the material in relation to the magnetic north pole. This can also be used to determine the changing position of continents in relation to the pole. In the months following Vine and Matthews' publication, however, the paper's significance was largely ignored with an American geologist dismissing its hypotheses as being "rather ridiculous". And Vine's paper remained in a state of limbo for the next 18 months.[17]

Canadian geologist John Tuzo Wilson proposed that volcanic island chains—such as the Hawaiian islands, for instance—were a result of a plate moving over a fixed "hotspot" in the Earth's mantle. As the plate continued its motion, each successive area over the "hotspot" would form a

[13] Frederick J. Vine, "Reversals of Fortune" in *Plate Tectonics: An Insider's History of the Modern Theory of the Earth*. Naomi Oreskes, (ed). (Boulder, Westview Press, 2003) p. 49.

[14] Vine, "Reversals of Fortune", pp. 50–51.

[15] Vine, "Reversals of Fortune", pp. 52, 53, 55.

[16] Vine, "Reversals of Fortune", p. 57.

[17] H.W. Menard, *The Ocean of Truth: A Personal History of Global Tectonics*. (Princeton, Princeton University Press, 1986) pp. 219–221.

volcanic island which would then become dormant as the plate moved on. In this way, a chain of volcanic islands would be created, the most active island indicating the position of the "hotspot" and the chain itself showing the direction of plate movement. Wilson's hypothesis provided a compelling explanation for volcanoes that did not appear to be confined to plate boundaries. Two years after this controversial proposition, Wilson then suggested that, in addition to the ocean ridges which was where crust appeared to be forming as well as trenches where crust was subducted, there was another form of boundary between plates that linked these two boundaries. He believed that this third boundary could "transform" into a fault. Unlike the other plate boundaries, transform faults were sites where crust was neither created nor destroyed but were regions where plates slid past each other.[18]

Both the San Andreas Fault and, to the north of it, the Queen Charlotte Islands fault system seemed to substantiate Wilson's proposition but his work also predicted that there would be a ridge between the two faults located off the coast of the north-western United States which he called the Juan de Fuca Ridge. As Frederick Vine recollected, Wilson was in the process of explaining this to him and Hess when Hess pointed out that the area had already been subject to a "detailed magnetic survey". The resulting map showed that, in the area of Wilson's hypothesised ridge, there were the same magnetic "stripes" characteristic of mid-ocean ridges. Even more significantly, these stripes displayed a mirror-image "symmetry" on either side of the ridge. This implied that, as magma welled up on both sides of the ridge, it cooled with the molten rock magnetised in the direction of the prevailing magnetic north. This was a concrete indication of seafloor spreading.[19] Wilson's work, thus, was able to provide the validation that Vine's hypothesis required, appeasing geologists who, while leery of Vine's theory located within the notion of seafloor spreading, were quite familiar with fault systems and so more inclined to take Wilson seriously.[20]

Final confirmation came with the use of core samples. Part of Vine's hypothesis related to the periodic reversal of the Earth's magnetic poles which was a relatively new discovery. In 1929 and 1936, there had been

[18] "J. Tuzo Wilson: Discovering transforms and hotspots" https://pubs.usgs.gov/gip/dynamic/Wilson.html, accessed 11 January 2017.

[19] Vine, "Reversals of Fortune", pp. 59–60.

[20] Menard, *The Ocean of Truth*, p. 245.

research pointing to the likelihood of magnetic field reversals. It was not until the 1950s, however, when J. Hospers—also studying at Cambridge University—wrote a paper "Reversals of the Main Geomagnetic Field" which was based on his study of rock samples obtained from the lava stratigraphic record in Iceland. The rock layers revealed alternating periods of "normal" and "reverse" polarity of the Earth's magnetic field. As the molten lava erupted onto the Earth's surface, elements within it were magnetised according to the prevailing polarity. This demonstrated that, throughout the Earth's history—as preserved in the stratigraphic record—the magnetic field "flipped" regularly with magnetic north becoming magnetic south and vice versa.[21] Core samples demonstrated that this pattern of periodic shifts in the magnetic field went back 3.6 million years.[22]

As palaeomagnetism began providing supporting evidence for plate tectonics, similar insights were derived from seismology. As seismic waves travel through media of different density at different velocities, this suggested that the Earth was composed of more than a solid iron core and a rocky outer crust than had initially been supposed.[23] By measuring the movement of waves, seismologists are able to create a three-dimensional map of the Earth's interior. The science was given impetus by the nuclear age. American nuclear tests taking place on Pacific islands in the mid-twentieth century—the most notorious of which was that at Bikini Atoll in 1954—generated seismic waves that could be used to study the Earth's interior. Geophysicists were no longer restricted to waiting for earthquakes.[24]

At the Lamont Geological Observatory, the geologists involved in the area of seismology were only too aware of the strides that were occurring in palaeomagnetism. Lynn Sykes was able to provide seismological proof of Wilson's transform faults. This then raised a further question. Now that seafloor spreading had been largely accepted, with the understanding that new crust was created at the ridges, how exactly was all this additional crust being accommodated? The Earth was either expanding or the crust was being "lost" in some manner. Again, geologists were divided, some

[21] Neil D. Opdyke, "The Birth of Plate Tectonics" in *Plate Tectonics: An Insider's History of the Modern Theory of the Earth*. Naomi Oreskes, (ed). (Boulder, Westview Press, 2003) pp. 95, 97, 98.

[22] Opdyke, "The Birth of Plate Tectonics", pp. 98–99, 101–102, 104–106.

[23] David Whitehouse, *Journey to the Centre of the Earth: A Scientific Exploration into the Heart of Our Planet*. (London, Weidenfeld and Nicholson, 2016) pp. 126–128.

[24] Whitehouse, *Journey to the Centre of the Earth*, pp. 129–130.

believing that the crust was being lost beneath continents whilst others (such as Henry Hess) believed that the crust disappeared into the interior at the deep-sea trenches. Jack Oliver at the observatory and Bryan Isacks were running a seismograph system in the South Pacific Polynesian islands in an area that had been characterised by deep earthquakes. Oliver makes clear that the research was "inductive"; it was not motivated by substantiating plate tectonics however their findings would solve the mystery of the disappearing of the Earth's crust. The region beneath the island arc up to 720 kilometres below the surface conducted seismic waves very efficiently and was the origin of the deep earthquakes. This was evidence for the subduction of the Earth's crust as it travelled into the mantle.[25] Palaeomagnetism had shown the birth of the Earth's crust; seismology had shown its death.

By the late 1960s plate tectonics was indeed a theory whose time had come. Computer programs and calculations would be used to reconstruct the Earth's history. At Cambridge, J.E. Everett and A.G. Smith used a computer model to accurately reconstruct the fit between South Africa and South America, between West Africa and Brazil and between Africa and North America (which was joined to Europe). While reconstructions remain open to refinement and debate, computer modelling substantiated Du Toit's extensive fieldwork and the technology of conflict continued its association with plate tectonics.[26]

The Supercontinent Cycle

In *Our Wandering Continents*, concerned as Alex Du Toit was with persuading his detractors of drift theory, it would have been somewhat incredulous to suggest that Pangaea—of which Gondwana was initially part—may have simply been the latest in a series of supercontinents. He does, however, suggest the tantalising possibility that continental movement was "an inherent property of the crust and as having operated throughout geological time and over the entire globe". The evidence for

[25] Jack Oliver, "Earthquake Seismology in the Plate Tectonics Revolution" in *Plate Tectonics: An Insider's History of the Modern Theory of the Earth* Naomi Oreskes, (ed). (Boulder, Westview Press, 2003) pp. 159–160.

[26] A. Hallam, *A Revolution in the Earth Sciences: From Continental Drift to Plate Tectonics*. (Oxford, Oxford University Press, 1973) pp. 68, 78–80.

this, however, was elusive and "fragmentary".[27] In this, too, Du Toit would be proved prescient.

Pangaea was not the first supercontinent, just the most recent. Periodically, over hundreds of millennia, continents come together to form supercontinents which subsequently rift apart. This has been termed the Supercontinent Cycle and is uniformitarianism on a grand scale. Determining the geological history of these supercontinents, like much of geology, has practical applications as well. The movement of continents affects ocean currents which plays a crucial role in climate. In addition, of particular economic benefit, it is at the continental shelves that oil deposits may be located.[28] The significance of the grand theory of plate tectonics is that, not only does it allow for an understanding of the history of Earth processes, the constant movement of the plates also enables the prediction of future processes, creating a dramatic break from earlier geological theories with their focus on the past. Yet the past was the important means by which the present (and future) was understood.

Professor John Joly was a geophysicist at Trinity College, Dublin. As with many geologists in the late nineteenth century, Joly was involved in attempting to determine the age of the Earth. Initially reaching a figure of 89 million years, he later realised the importance of radioactivity in dating the age of rocks as the radioactive elements within rocks decay at a predictable rate. His major contribution to geology came, however, in this same application of radioactivity to continental rifting. According to Joly, in a lecture presented at Oxford University in 1924, rocks within the Earth's crust radiate heat as a result of the decay of their radioactive components. This interior heat accumulates as the Earth's crust provides a layer of insulation. Over many millions of years, this heat would lead to the melting of the rock, resulting in the fracturing of the continents that lay above them. This would culminate in volcanic eruptions on an incredible scale and possibly lead to the movement of continents. Evidence for this volcanism could be seen in the Deccan Traps in India as well as the Siberian Traps—the latter due to the fragmentation of Pangaea. Joly's proposition still remains the most apposite for the rifting of supercontinents.[29]

[27] T.W. Gevers, "The Life and Work of Dr. Alex. L. Du Toit"—Alex L. Du Toit Memorial Lectures No. 1. In *The Geological Society of South Africa: Annexure to Volume LII* (Johannesburg, The Geological Society of South Africa, The South African Association for the Advancement of Science, The South African Geological Society, 1949) p. 89.

[28] Nield, *Supercontinent,* pp. xiii, 12.

[29] Nield, *Supercontinent,* pp. 174–175, 179, 181–183.

The periodic build-up of interior heat thus suggests a periodic formation and break-up of supercontinents however the evidence for the supercontinents that preceded Pangaea is meagre. The evidence can only be located in ancient rock, the metamorphosed, deformed and often inaccessible remnants that have escaped subduction. According to geologist Trond Torsvik, trying to reconstruct past supercontinents "resemble a jigsaw puzzle, where we must contend with missing and faulty pieces and have misplaced the picture on the box". Moreover, the deformation experienced by these ancient records—as well as their age—means that there is a dearth of fossil evidence that allows for correlation. Even with the aid of sophisticated computer programming, geologists are still in disagreement of the sequence of supercontinents and, as they move further back into time, it becomes even more complicated.[30]

Here, again, however important evidence was derived from Africa. The first supercontinent is believed to be Ur—a biblical reference that once again highlights the interaction between history, mythology and geology as well as the very human desire to contextualise the distant unknown within the familiar. Ur is believed to have existed about three billion years ago and was already experiencing the geological processes that characterise uniformitarianism—erosion, deposition and glaciation. It comprised many of the landmasses that would later make up Gondwana and some of the most ancient rocks in the world are found here. Evidence for these processes have been found in South Africa in the ancient Witwatersrand and Pongola basins that have sedimentary rock indicating the movement of water and glaciation. It was also the supercontinent that saw the birth of simple life.[31] In vindication of Du Toit, "Africa forms the key".

[30] Nield, *Supercontinent,* pp. 191, 201.
[31] Nield, *Supercontinent,* pp. 229–230.

Bibliography

Archival Sources and Newspaper Articles

Alex L. Du Toit Papers—BC 722, University of Cape Town Libraries.

Smuts Papers Vol 34, No 122: Letter from IB Pole Evans to Smuts, 30 March 1927, University of Cape Town Libraries.

Council for Geoscience—Library Information Centre: Address by Alex Du Toit to Special General meeting of the GSSA, 27 April, 1927, reprinted in *IGC*, Vol 25, No 1, 1929.

Council for Geoscience—Library Information Centre: B0301693—Alex Du Toit: "The Aftermath of the Fifteenth International Geological Congress: An appreciation from South Africa" in *Economic Geology*, Vol 28.

Council for Geoscience—Library Information Centre: *International Geological Congress: Compte Rendu of the XV Session, South Africa, 1929—Vol II: Scientific Communication*, Pretoria, Wallachs' Ltd, 1930.

Council for Geoscience—Library Information Centre: Letter by Cesar Rubio to A.L. Hall, 28 March, 1926 reprinted in *IGC*, Vol 25, No 1, 1929.

Geological Society of South Africa (GSSA): "International Geological Congress XV Session, South Africa, 1929: Guidebook", Pretoria, Wallachs Ltd, 1929.

"£500 City Gift to War Fund" in *The Rand Daily Mail*, Saturday, 3 August 1940.

"Another £600 for Raid Fund" in *The Rand Daily Mail*, Thursday 17 October 1940.

Bright, L., "Men, Women, Events: A Daily Causerie" in *Rand Daily Mail*, Saturday, 22 May 1948, http://phw02.newbank.com/cache/arhb/fullsize/pl_010272016_0606_32763_126.pdf, accessed 27 October 2016.

S. Chetty, *'Africa Forms the Key'*, Cambridge Imperial and Post-Colonial Studies, https://doi.org/10.1007/978-3-030-52711-2

"The British Association at Victoria Falls: Opening of the Bridge" in *The Times*, September 13, 1905, Issue 37811, p. 9, London, Times Newspapers Limited, Gale Document Number CS151187757.

"The British Association in Cape Town" in *The Times*, August 17, 1905, Issue 37788, p. 4, London, Times Newspapers Limited, Gale Document Number CS84078864.

"The British Association in South Africa" in *The Times*, August 16, 1905, Issue 37787, p. 5, London, Times Newspapers Limited, Gale Document Number CS84078864.

"The British Association in South Africa" in *The Times*, August 18, 1905, Issue 37790, p. 4, London, Times Newspapers Limited, Gale Document Number CS67432723.

"The British Association in South Africa" in *The Times*, September 16, 1905, Issue 37814, p. 11, London, Times Newspapers Limited, Gale Document Number CS184742192.

"The British Association in the Transvaal" in *The Times*, September 1, 1905, Issue 37801, p. 8, London, Times Newspapers Limited, Gale Document Number CS134410529.

"The British Association in the Transvaal" in *The Times*, September 2, 1905, Issue 37802, p. 9, London, Times Newspapers Limited, Gale Document Number CS151449890.

"The British Association: London Meeting Opened" in *The Times*, September 24, 1929, Issue 45936, p. 17, London, Times Newspapers Limited, Gale Document Number CS117646648.

"The British Association Opening of Norwich Meeting: Professor Watts on the Chain of Life" in *The Times*, September 5, 1935, Issue 47161, p. 6, London, Times Newspapers Limited, Gale Document Number CS100871461.

"Discussion on a Lost Continent: Evidence from Fish" in *The Times*, August 5, 1929, Issue 45273, p. 9, London, Times Newspapers Limited, Gale Document Number CS15272869.

"Fund in Memory of SA Scientist" in *Rand Daily Mail*, Saturday, 24 July 1948, http://phw02.newbank.com/cache/arhb/fullsize/pl_010272016_0557_01373_303.pdf, accessed 27 October 2016.

McKie, R., "Scott of the Antarctic: the lies that doomed his race to the pole" in *The Guardian*, Saturday, 24 September, 2011, http://www.theguardian.com/uk/2011/sep/24/scott-antarctic-lies-race-pole, accessed 17 October 2016.

Robson, D., "The Scott expedition: How science gained the pole position" in *The Telegraph*, 21 June 2011, http://www.telegraph.co.uk/news/science/science-news/8587530/The-Scott-expedition-how-science-gained-the-pole-position.html, accessed 17 October 2016.

"The Royal Society: New Fellows Elected" in *The Times*, March 19, 1943, Issue 49498, p. 2, London, Times Newspapers Limited, Gale Document Number CS34815091.

"SA Scientist to be Honoured by Series of Lectures" in *Rand Daily Mail*, Thursday, 20 October 1949. http://phw02.newbank.com/cache/arhb/fullsize/pl_010272016_0550_15136_454.pdf, accessed 27 October 2016.

"Science and the Empire: Sir R. Gregory on Research" in *The Times*, December 4, 1929, Issue 45377, p. 16, London, Times Newspapers Limited, Gale Document Number CS268902788.

"Science in Africa" in *The Times*, July 23, 1929, Issue 45262, p. 15, London, Times Newspapers Limited, Gale Document Number CS251994359.

"Scientists in S. Africa" in *The Times*, July 23, 1929, Issue 45262, p. 14, London, Times Newspapers Limited, Gale Document Number CS67301649.

Alex Du Toit: Books and Articles

"Arthur William Rogers" in *South African Journal of Geology*, Vol 49, pp. 291–304, 1946, http://www.sabinet.co.za, AJA10120750_1936, accessed 13 October 2016.

"Borehole Water Supplies in the Union of South Africa" in *Civil Engineering*, Vol 26, pp. 74–133, 1928, http://www.sabinet.co.za Doc ATA10212019_21628, accessed 13 October 2016.

"The Carboniferous Glaciation of South Africa" in *Transactions of the Geological Society of South Africa*, Johannesburg: Geological Society of South Africa, 1921.

"The Contribution of South Africa to the Principles of Geology" in *South African Journal of Science*, Vol 21, pp. 52–78, November 1924, http://www.sabinet.co.za Doc AJA00382353_2525, accessed 13 October 2016.

A Geological Comparison of South America with South Africa, Washington, Carnegie Institution of Washington, 1927.

The Geology of South Africa, London and Edinburgh, Oliver and Boyd, 1926.

"The Geology of Underground Water Supply with Special Reference to South Africa" in *Civil Engineering*, Vol 1, pp. 8–31, 1913, http://www.sabinet.co.za Doc AJA10212019_185214, accessed 13 October 2016.

"Land Connections between the Other Continents and South Africa in the Past", Public Lecture given on 15 July, 1921. Published in *South African Journal of Science*, 18, 120–140.

Our Wandering Continents: An Hypothesis of Continental Drifting, London and Edinburgh, Oliver and Boyd, 1937.

Physical Geography for South African Schools, Cambridge, Cambridge University Press, 1926.

"Reply on Underground Water Supply", Fourth Meeting, 11th June 1913, Sabinet Doc AJA10212019-185214, accessed 13 October 2016.

"Some Reminiscences" in *Transactions of the Geological Society of South Africa*, Jubilee Volume, 1946 reprinted in T.W. Gevers, "The Life and Work of Dr. Alex. L. Du Toit"—Alex L. Du Toit Memorial Lectures No. 1. In *The Geological Society of South Africa: Annexure to Volume LII*, Johannesburg, The Geological Society of South Africa, The South African Association for the Advancement of Science, The South African Geological Society, 1949.

Secondary Sources: Books, Journal Articles and Chapters and Online Sources

A.G. "Obituary: Alexander Logie Du Toit, F.R.S." in *the South African Archaeological Bulletin*, Vol 3, No 9, March 1948.

Aldrich, M.L., "Taylor, Frank Bursley" in *Complete Dictionary of Scientific Biography* (Charles Scribner's Sons: 2008) taken from http://www.encyclopedia.com/people/science-and-technology/geology-and-oceanography-biographies/frank-bursley-taylor, accessed 7 October 2016.

Alvarez, A., *A Most Improbable Journey: A Big History of Our Planet and Ourselves*, New York, W.W. Norton and Company, 2017.

Anderson, R.D., *Education and the Scottish People, 1750–1918*, Oxford, Oxford University Press, 1995.

Anker, P., "The Politics of Ecology in South Africa on the Radical Left" in *Journal of the History of Biology*, Vol 37, No 2, Summer 2004.

"Arthur P. Coleman—Biographical Sketch" Victoria University Library Special Collections, University of Toronto http://library.vicu.utoronto.ca/collections/special_collections/f7_arthur_p_coleman, accessed 15 March 2017.

"The Basin", Great Artesian Basin Coordinating Committee, http://www.gabcc.gov.au/basin, accessed 18 January 2017.

"Basil Schonland (1896–1972)" http://www.thepresidency.gov.za/national-orders/recipient/basil-schonland-1896-1972, accessed 5 March 2018.

Beinart, W., *The Rise of Conservation in South Africa: Settlers, Livestock, and the Environment 1770–1950*, Oxford, Oxford University Press, 2003.

Benedetti, J.J., "Memorial to Horacio J. Harrington, 1910–1973" http://rock.geosociety.org/pub/Memorials/v07/Harrington-HJ.pdf, accessed 1 September 2017.

Blackwelder, E., "Bailey Willis: 1857–1949", Washington D.C.: National Academy of Sciences, 1961, www.nasonline.org/publications/biographical-memoirs/memoir-pdfs/willis-bailey.pdf, accessed 11 September 2017.

Bonner, P., "Africa is Seldom What It Seems" in *A Search for Origins: Science, History and South Africa's 'Cradle of Humankind'*, Philip Bonner, Amanda Esterhuysen and Trefor Jenkins, (eds), Johannesburg, Wits University Press, 2007.

Boucher, M., *The University of the Cape of Good Hope and the University of South Africa, 1873–1946: A Study in National and Imperial Perspective*, Pretoria, The Government Printer, 1974.

Broken Hill—Olary Geology and Geophysics Field Course http://www.geosci.usyd.edu.au/users/prey/FieldTrips/BrokenHillOlary/Mapping.html, accessed 17 January 2016.

Brown, C.S., *Big History: From the Big Bang to the Present*, New York, The New Press, 2012.

Cantrill, D.J. and Poole, I., *The Vegetation of Antarctica through Geological Time*, Cambridge, Cambridge University Press, 2012.

Cawthorn, R.G., H.V. Eales, H.V., Walrave, F., Uken, R. and Watkeys, M.K., "The Bushveld Complex" in *The Geology of South Africa*, M.R. Johnson, C.R. Anhaeusser and R.J. Thomas, (eds), Johannesburg and Pretoria, The Geological Society of South Africa and the Council for Geoscience, 2006.

Chakrabarti, P., "Gondwana and the Politics of the Deep Past" in *Past and Present*, 242:1, February 2019.

Chetty, S., "David Draper: The Making of a South African Geologist" in *Historia*, Vol 62, No 2, November 2018.

Christian, D., *Origin Story: A Big History of Everything*, Allen Lane, 2018.

Cloud, P., "The Second Flowering of Geology" in *Bulletin of the American Academy of Arts and Sciences*, Vol 36, No. 3 (Dec, 1982).

Crowell, W.U., "The Evolution of South African Control over South West Africa (Namibia)", St John's University, Unpublished PhD Dissertation, 1975.

Cullis, S., "The Heroic Antarctic Explorers and the South African War of 1899–1902" in *The South African Military History Society* Newsletter No 456, August 2017.

Dart, R. and Craig, D., *Adventures with the Missing Link*, London, Hamish Hamilton Ltd, 1959.

Davenport, J., *Digging Deep: A History of Mining in South Africa*, Johannesburg, Jonathan Ball Publishers, 2013.

Desmond, A. and James Moore, J., *Darwin*, London, Penguin Books, 2009.

De Wit, M.J., De Brito Neves, B.B., Trouw, R.A.J. and Pankhurst, R.J., "Pre-Cenozoic Correlations across the South Atlantic Region: 'the ties that bind'" in *West Gondwana: Pre-Cenozoic Correlations Across the South Atlantic Region*, London, The Geological Society of London 2008), downloaded from http://sp.lyellcollection.org, accessed 2 October 2016.

"Dr David Draper (1849–1929)" in *A Century of Geological Endeavour in Southern Africa, 1895–1995*, C.R. Anhaeusser, (ed), Linden, The Geological Society of South Africa, 1997.

Dubow, S., *A Commonwealth of Knowledge: Science, Sensibility and White South Africa, 1820–2000*, Oxford, Oxford University Press, 2007.

Dubow, S., "A Commonwealth of Science: the British Association in South Africa, 1905 and 1929" in *Science and Society in Southern Africa*, Saul Dubow (ed), Manchester, Manchester University Press, 2000.

Dubow, S., "Global Science, National Horizons: South Africa in Deep Time and Space" in *The Historical Journal*, Cambridge University Press, 2020.

Dubow, S., *Henri Breuil and the Imagination of Prehistory: 'Mixing up Rubble, Trouble and Stratification'* in South African Archaeological Society Goodwin Series 12: 2019.

Dubow, S., "Human Origins, Racial Typology and the Other Raymond Dart" in *African Studies*, 55:1, 1996.

Dubow, S., *Scientific Racism in Modern South Africa*, Cambridge, Cambridge University Press, 1995.

Dubow, S., "White South Africa and the South Africanisation of Science: Humankind or Kinds of Humans?" in *A Search for Origins: Science, History and South Africa's 'Cradle of Humankind'*, Philip Bonner, Amanda Esterhuysen and Trefor Jenkins, (eds), Johannesburg, Wits University Press, 2007.

Ellenberger, F., "The First International Geological Congress (1878)" http://iugs.org/uploads/images/PDF/1st%20IGC.pdf, accessed 2 November 2016.

Fagan, B., *Cro-Magnon: How the Ice Age gave Birth to the First Modern Humans*, New York, Bloomsbury Press, 2010.

Fermor, L.L., "Thomas Henry Holland, 1868–1947" in *Obituary Notices of Fellows of the Royal Society*, Vol 6, No 17, Nov 1948.

Fortey, R., *The Hidden Landscape: A Journey into the Geological Past*, London, Bodley Head, 2010.

Frankel, H.R., *The Continental Drift Controversy*, Vols 1–4, Cambridge, Cambridge University Press.

Galbraith, J.S., *Crown and Charter: The Early Years of the British South Africa Company*, Berkeley and Los Angeles, University of California Press, 1974.

Gascoigne, J., *Science in the Service of Empire: Joseph Banks, the British State and the Uses of Science in the Age of Revolution*, Cambridge, Cambridge University Press, 2010.

http://www.geolsoc.org.uk/history, accessed 12 December 2014.

http://www.geolsoc.org.uk/About/Awards-Grants-and-Bursaries/Society-Awards/Murchison-Medal, accessed 1 February 2016.

Gevers, T.W., "The Life and Work of Dr. Alex. L. Du Toit"—Alex L. Du Toit Memorial Lectures No. 1. In *The Geological Society of South Africa: Annexure to Volume LII*, Johannesburg, The Geological Society of South Africa, The South African Association for the Advancement of Science, The South African Geological Society, 1949.

Giliomee, G., *The Afrikaners: Biography of a People*, Cape Town, Tafelberg Publishers, 2003.

https://global.britannica.com/science/charnockite, accessed 23 January 2017.

Gohau, G., *A History of Geology*, Albert V. Carozzi and Marguerite Carozzi, (translators), New Brunswick and London, Rutgers University Press, 1990.

Gould, S.J., *Time's Arrow, Time's Cycle: Myth and Metaphor in the Discovery of Geological Time*, Penguin Books, 1991.

Gould, S.J., *Wonderful Life: The Burgess Shale and the Nature of History*, New York, WW Norton and Company, 1990.

Gray, E. and Gray, J., *A History of the Discovery of the Witwatersrand Goldfields*, Johannesburg, Sholto Douglas and Company, 1940.

Greene, M.T., *Alfred Wegener: Science, Exploration, and the Theory of Continental Drift*, Baltimore, Johns Hopkins University Press, 2015.

Grotzinger, J. and Jordan, T.H., *Understanding Earth (Sixth Edition)*, New York: W.H. Freeman and Company, 2010.

Hallam, A., *Great Geological Controversies*, Oxford, Oxford University Press, 1992.

Hallam, A., *A Revolution in the Earth Sciences: From Continental Drift to Plate Tectonics*, Oxford, Oxford University Press, 1973.

Hancock, W.K., *Smuts 2: The Fields of Force, 1919–1950*, Cambridge, Cambridge University Press, 1968.

Haughton, S.H., "Alexander Logie Du Toit, 1878–1948" in *Obituary Notices of Fellows of the Royal Society*, Vol 6, No 18, November 1949.

Haughton, S.H., 'South African Geology' in *A History of Scientific Endeavour in South Africa: A collection of essays published on the occasion of the Centenary of the Royal Society of South Africa*, A.C. Brown, (ed), Cape Town, The Royal Society of South Africa, 1977.

Heyman, R.D., "C.T. Loram: A South African Liberal in Race Relations" in *The International Journal of African Historical Studies*, Vol 5, No1, 1972.

Hofmann, T., "Milestones of a Life beyond the Geosciences" in *The Face of the Earth: the Legacy of Eduard Suess*, T. Hofmann, G. Bloschl, L. Lammerhuber, W.E. Piller, A.M. Celal Sengor, (eds), Munich, European Geosciences Union, 2014.

Hofmeyr, J.H., "Africa and Science" in Science, Vol LXX, No. 1812, Friday, September 20, 1929.

"The International Geological Congress (A Brief History), http://iugs.org/uploads/images/PDF/A%20Brief%20History.pdf, accessed 2 November 2016.

https://ipfs.io/ipfs/QmXoypizjW3WknFiJnKLwHCnL72vedxjQkDDP1mXWo6uco/wiki/Alexander_du_Toit.html, accessed 27 July 2018.

"J. Tuzo Wilson: Discovering transforms and hotspots" https://pubs.usgs.gov/gip/dynamic/Wilson.html, accessed 11 January 2017.

James, H.L., *Harry Hammond Hess, 1906–1969: A Biographical Memoir*, (Washington D.C., National Academy of Sciences, 1973.

Johnson, M.R., Van Vuuren, C.J., Visser, J.N.J., Cole, D.I., Wickens, H. De V., Christie, A.D.M., Roberts, D.L. and Brandl, G., "Sedimentary Rocks of the

Karoo Supergroup" in *The Geology of South Africa*, M.R. Johnson, C.R. Anhaeusser, R.J. Thomas, (eds), Johannesburg and Pretoria, The Geological Society of South Africa and the Council for Geoscience, 2006.

Jones, N., "Minerals yield signs of early plate tectonics" in *Nature*, 26 November 2008. http://www.nature.com/news/2008/081126/full/news.2008.1256.html, accessed 4 May 2017.

Karrstrom, E.J., *Eighteen Years in South Africa: A Swedish Gold-Digger's Account of His Adventures in the Land of Gold (1877–1896)* I. Rudner, (ed), I and J Rudner, (translators), Cape Town, Africana Publishers, 2013.

Klutschak, H.W. "Ein Besuch auf Sud Georgien", R.S. Boumphrey, translator, in *British Antarctic Survey Bulletin*, 12, 1881.

Knopf, A., "Charles Schuchert: 1858–1942", Washington D.C., National Academy of Sciences, 1952, www.nasonline.org/publications/biographical-memoirs/memoir-pdfs/schuchert-charles.pdf, accessed 12 September 2017.

Kuljian, C., *Darwin's Hunch: Science, Race and the Search for Human Origins*, Johannesburg, Jacana Media, 2016.

Leasor, J., *Rhodes and Barnato: The Premier and the Prancer*, London, Leo Cooper, 1997.

Leverett, F., "Memorial to Frank Bursley Taylor", *Proceedings of the Geological Society of America for 1938* (1939).

Levin, G., 'Andrew Geddes Bain' in 'South African Geological Association (1888–1890)' reprinted in *A Century of Geological Endeavour in Southern Africa, 1895–1995*, C.R. Anhaeusser, (ed), Linden, The Geological Society of South Africa, 1997.

Leviton, A.E. and Aldrich, M.L., "The Impact of Travels on Scientific Knowledge: William Thomas Blanford, Henry Francis Blanford, and the Geological Survey of India, 1851–1889" in *Proceedings of the California Academy of Sciences*, Vol 55, Supplement II, No 9, November 19, 2004.

Lewsen, P., *John X. Merriman: Paradoxical South African Statesman*, Johannesburg, AD Donker, 1982.

Livingstone, D.N., *Putting Science in its Place: Geographies of Scientific Knowledge*, Chicago and London, The University of Chicago Press, 2003.

Ludwig-Maximilians-Universitaet Muenchen (LMU) "How diamond-bearing kimberlites reach the surface of Earth: Acidification provides the thrust", 26 January 2012. http://www.sciencedaily.com/releases/2012/01/120123094523.htm, accessed 9 December 2014.

Machens, E.W., *Platinum, Gold and Diamonds: The adventure of Hans Merensky's discoveries*, Idette Noome, (translator), Pretoria, Protea Book House, 2013.

Mackie, J.D., *The University of Glasgow, 1451–1951: A Short History*, Glasgow, Jackson, Son and Company, 1954.

MacRae, C., *Life Etched in Stone: Fossils of South Africa*, Johannesburg, The Geological Society of South Africa, 1999.

Maldonada, A., Dalziel, I.W.D. and Leat, P.T., Corrigendum to "The global relevance of the Scotia Arc: An introduction" in *global and Planetary Change*, vol 133, October 2015, p. 378. Taken from http://www.sciencedirect.com/science/article/pii/S0921818114001404, accessed 6 December 2016.

Marks, S., "War and Union, 1899–1910" in *The Cambridge History of South Africa, Volume 2: 1885–1994*, R. Ross, A.K. Melger and B Nasson, (eds), Cambridge, Cambridge University Press, 2011.

Mason, R., "Stripes on the Sea Floor" in *Plate Tectonics: An Insider's History of the Modern Theory of the Earth.* Naomi Oreskes, (ed), Boulder, Westview Press, 2003.

Mazarire, G.C., "Reflections on Pre-Colonial Zimbabwe, *c.* 850–1880s" in *Becoming Zimbabwe: A History from the Pre-colonial Period to 2008*, Brian Raftopoulos and Alois Mlambo, (eds), Harare, Weaver Press, 2009.

McPhee, J., *Annals of the Former World*, New York, Farrar, Straus and Giroux, 2000.

Menard, H.W., *The Ocean of Truth: A Personal History of Global Tectonics*, Princeton, Princeton University Press, 1986.

Naude, S.M. and Brown, A.C., 'The Growth of Scientific Institutions in South Africa' in *A History of Scientific Endeavour in South Africa: A collection of essays published on the occasion of the Centenary of the Royal Society of South Africa*, A.C. Brown, (ed), Cape Town, The Royal Society of South Africa, 1977.

Ndlovu-Gatsheni, S.J., "Mapping Cultural and Colonial Encounters, 1880s–1930s" in *Becoming Zimbabwe: A History from the Pre-colonial Period to 2008*, Brian Raftopoulos and Alois Mlambo, (eds), Harare, Weaver Press, 2009.

Neubauer, F., "Gondwana-land Goes Europe" in *Austrian Journal of Earth Sciences*, Vol 107/1, 2014.

Newton, A.R., Shone, R.W. and Booth, P.W.K., "The Cape Fold Belt" *The Geology of South Africa*, M.R. Johnson, C.R. Anhaeusser, R.J. Thomas, (eds), Johannesburg and Pretoria, The Geological Society of South Africa and the Council for Geoscience, 2006.

Nield, T., *Supercontinent: Ten Billion Years in the Life of Our Planet*, London, Granta Books, 2008.

Oliver, J., "Earthquake Seismology in the Plate Tectonics Revolution" in *Plate Tectonics: An Insider's History of the Modern Theory of the Earth.* Naomi Oreskes, (ed), Boulder, Westview Press, 2003.

Opdyke, N.D., "The Birth of Plate Tectonics" in *Plate Tectonics: An Insider's History of the Modern Theory of the Earth.* Naomi Oreskes, (ed), Boulder, Westview Press, 2003.

Oreskes, N., "From Continental Drift to Plate Tectonics" in *Plate Tectonics: An Insider's History of the Modern Theory of the Earth.* Naomi Oreskes, (ed), Boulder, Westview Press, 2003.

Oreskes, N., *The Rejection of Continental Drift: Theory and Method in American Earth Science*, New York, Oxford University Press, 1999.

The Oxford Companion to the Earth, P.L. Hancock and B.J. Skinner, (eds), Oxford, Oxford University Press, 2000.

Paton, A., *Hofmeyr*, Cape Town, Oxford University Press, 1964.

Pettijohn, F.J., "Rollin Thomas Chamberlin", Washington D.C.: National Academy of Sciences, 1970, http://www.nasonline.org/publications/biographical-memoirs/memoir-pdfs/chamberlin-rollin-t.pdf, accessed 14 March 2017.

http://www.paleophilatelie.eu/description/stamps/south_africa_1991.html, accessed 11 June 2020.

Phillips, H., *The University of Cape Town, 1918–1948: The Formative Years*, Cape Town, University of Cape Town Press, 1993.

Piller, W.E., "From Palaeontology and Stratigraphy to Earth System Science" in *The Face of the Earth: the Legacy of Eduard Suess* T. Hofmann, G. Bloschl, L. Lammerhuber, W.E. Piller, A.M. Celal Sengor, (eds), Munich, European Geosciences Union, 2014.

Plug, C., "The origin and early history of S_2A_3" in *Rudolf Marloth Brochure: Centenary Edition 1902–2002* Cornelis Plug, ed, Boordfontein: Forum Press, 2002, http://s2a3.org.za/joomla/files/archives/Marlth_Brochure_Centenary_Edition_2002.pdf, accessed 2 December 2016.

Plug, C., "Pole-Evans, Mr Illtyd Buller" in *S2A3 Biographical Database of Southern African Science*, http://www.s2a3.org.za/bio/Biograph_final.php?serial=2224, accessed 7 February 2018.

Plug, C. and Levin, G., 'South African Geological Association (1888–1890)' in *A Century of Geological Endeavour in Southern Africa, 1895–1995*, C.R. Anhaeusser, (ed), Linden, The Geological Society of South Africa, 1997.

Porter, R., *The Making of Geology: Earth Science in Britain, 1660–1815*, Cambridge, Cambridge University Press, 1977.

Quilty, P.G. and Banks, M.R., "Samuel Warren Carey 1911–2002". Australian Academy of Science, https://www.science.org.au/fellowship/fellows/biographical-memoirs/samuel-warren-carey-1911-2002, accessed 15 March 2017.

Rapela, C.W. and Depetris, P.J., "Geochemistry in Argentina: from pioneers to the present" in *Environmental Earth Sciences*, 75:524, 2016.

Repcheck, R., *The Man Who Found Time: James Hutton and the Discovery of the Earth's Anatomy*, New York, Basic Books, 2009.

"Report of the Nineteenth Annual Meeting of the South African Association for the Advancement of Science" in *The South African Journal of Science*, Vol XVIII, Johannesburg, South African Association for the Advancement of Science, 1922.

Review of "Theory of Continental Drift: A symposium on the origin and movement of land masses both inter-continental and intra-continental, as proposed by Alfred Wegener in Quarterly Journal of the Royal Meteorological Society",

Vol 55, Issue 229, January 1929, p. 102, http://onlinelibrary.wiley.com/doi/10.1002/qj.49705522923/pdf, accessed 11 September 2017.

"Researchers confirm the existence of a 'lost continent' under Mauritius", 31 January 2017. https://phys.org/news/2017-01-lost-continent-mauritius.html, accessed 4 May 2017.

"Richard E. Byrd: Byrd Antarctic Expedition II" http://www.south-pole.com/p0000108.htm, accessed 5 March 2018.

Robinson, M.F., *The Lost White Tribe: Explorers, Scientists, and the Theory that Changed a Continent*, New York, Oxford University Press, 2016.

Rodgers, J., "Chester Ray Longwell", Washington D.C.: National Academy of Sciences, 1970, http://www.nasonline.org/publications/biographical-memoirs/memoir-pdfs/longwell-chester.pdf, accessed 15 March 2017.

Rogers, A.W., *The Pioneers in South African Geology and their Work (Transactions of the Geological Society of South Africa, Annexure to Volume XXXIX)*, Geological Society of South Africa, 1937.

Ross, R., *A Concise History of South Africa, Second Edition*, Cape Town, Cambridge University Press, 1999, 2008.

Rudwick, M.J.S., *Bursting the Limits of Time: The Reconstruction of Geohistory in the Age of Revolution*, Chicago, University of Chicago Press, 2007.

Rudwick, M.J.S., *The Great Devonian Controversy: The Shaping of Scientific Knowledge among Gentlemanly Specialists*, Chicago, University of Chicago Press, 1988.

Rudwick, M.J.S., *The Meaning of Fossils: Episodes in the History of Palaeontology*, New York, Science History Publications, 1976.

Rudwick, M.J.S., *Roy Porter, Historian of Geology* in *History of Science*, xli, 1 September 2003.

Rudwick, M.J.S., *Worlds before Adam: The Reconstruction of Geohistory in the Age of Reform*, Chicago, University of Chicago Press, 2010.

Sagan, C., Druyan, A. and Soter, S, *Cosmos*, Gregory Andorfer and Rob McCain, producers, 1980.

Sagan, C., *The Demon-Haunted World: Science as a Candle in the Dark*, New York, Ballantine Books, 1997.

Sagan, C., *Pale Blue Dot: A Vision of the Human Future in Space*, New York: Random House, 1994.

Sengor, A.M.C., "Eduard Suess and the Origin of Modern Geology" in *The Face of the Earth: the Legacy of Eduard Suess*. T. Hofmann, G. Bloschl, L. Lammerhuber, W.E. Piller, A.M. Celal Sengor, (eds), Munich, European Geosciences Union, 2014.

Sengor, A.M.C., "Suess and the Dynamics of the Planet Earth" in *The Face of the Earth: the Legacy of Eduard Suess*, T. Hofmann, G. Bloschl, L. Lammerhuber, W.E. Piller, A.M. Celal Sengor, (eds), Munich, European Geosciences Union, 2014.

Shepard, F.P., "Memorial to Phillip H. Kuenen, 1902–1976", The Geological Society of America, 1978, ftp://rock.geosociety.org/pub/Memorials/v08/Kuenen-PH.pdf, p. 1, accessed 15 September 2017.

Skinner, B.J., Porter, S.C. Jeffrey Park, J., *Dynamic Earth*, Hoboken, John Wiley and Sons, 2013.

Smuts, J.C., *Holism and Evolution*, New York, The Macmillan Company, 1926.

Steyn, R., *Jan Smuts: Unafraid of Greatness*, Johannesburg and Cape Town, Jonathan Ball Publishers, 2016.

Stilwell, J.D. and Long, J.A., *Frozen in time: Prehistorical Life in Antarctica*, Collingwood, Csiro Publishing, 2011.

Stone, P., "Geological exploration of South Atlantic islands and its contributions to the continental drift debate of the early 20th century" in *Proceedings of the Geologists' Association*, 2015, Vol 126, 266–281, taken from http://nora.nerc.ac.uk/510724/1/SouthAtlanticIslands-pga.pdf, accessed 12 October 2016.

Stubblefield, C.J., "Darashaw Nosherwan Wadia, 1883–1969" in *Biographical Memoirs of Fellows of the Royal Society*, 1 November 1970, Vol 16, p. 544. http://rsbm.royalsocietypublishing.org/content/7/19/264, accessed 5 September 2017.

Suess, E., *The Face of the Earth Vol I*, Oxford, Clarendon Press, 1904, reprinted in *The Face of the Earth: the Legacy of Eduard Suess.* T. Hofmann, G. Bloschl, L. Lammerhuber, W.E. Piller, A.M. Celal Sengor, (eds), Munich, European Geosciences Union, 2014.

Taylor, F.B., "Bearing of the Tertiary Mountain Belt on the Origin of the Earth's Plan" in *Bulletin of the Geological Society of America*, Vol 21, pp. 179–226, Pl. 4, June 3, 1910.

Thamm, A.G. and Johnson, M.R., "The Cape Supergroup" in *The Geology of South Africa.* M.R. Johnson, C.R. Anhaeusser and R.J. Thomas, (eds), Johannesburg and Pretoria, The Geological Society of South Africa and the Council for Geoscience.

Thomas, H.H., "Birbal Sahni, 1891–1949" in *Biographical Memoirs of Fellows of the Royal Society*, 1 November 1950, Vol 7, Issue 19, p. 270, http://rsbm.royalsocietypublishing.org/content/7/19/264, accessed 5 September 2017.

Tobias, P.V., "A Century of Research in Human Biology and Palaeo-anthropology in Southern Africa" in *A History of Scientific Endeavour in South Africa: A collection of essays published on the occasion of the Centenary of the Royal Society of South Africa*, A.C. Brown, (ed), Cape Town: The Royal Society of South Africa, 1977.

Tobias, P.V., "The Story of Sterkfontein Since 1895" in *A Search for Origins: Science, History and South Africa's 'Cradle of Humankind'*, Philip Bonner, Amanda Esterhuysen and Trefor Jenkins, (eds), Johannesburg: Wits University Press, 2007.

Trapido, S., "Imperialism, Settler Identities, and Colonial Capitalism: the Hundred Year Origins of the 1899 South African War" in *The Cambridge History of South Africa, Volume 2: 1885–1994*, R. Ross, A.K. Melger and B Nasson, (eds), Cambridge, Cambridge University Press, 2011.

University of Cincinnati "Memorial: Kenneth Edward Caster (1908–1992)" http://homepages.uc.edu/~huffwd/Department_History/Kenneth_E_Caster/Caster_Memorial.htm, accessed 1 September 2017.

Vai, G.B., "Giovanni Capellini and the origin of the International Geological Congress" in *Episodes*, Vol 25, No 4, December 2002, http://iugs.org/uploads/images/PDF/Capelllini.pdf, accessed 2 November 2016.

Vai, G.B., "The Second International Geological Congress, Bologna, 1881" in *Episodes* Vol 27, No1. http://iugs.org/uploads/images/PDF/2nd%20IGC.pdf, accessed 2 November 2016.

Van der Waag, I., *A Military History of Modern South Africa*, Johannesburg and Cape Town, Jonathan Ball Publishers, 2015.

Van Everdingen, E., "The Snellius Expedition" in *ICES Journal of Marine Science*, 1930, https://academic.oup.com/icesjms/article-pdf/5/3/320/1825039/5-3-320.pdf, accessed 15 September 2017.

Vine, F.J., "Reversals of Fortune" in *Plate Tectonics: An Insider's History of the Modern Theory of the Earth.* Naomi Oreskes, (ed), Boulder, Westview Press, 2003.

Warhurst, P.R., "Smuts and Africa: a study in sub-imperialism" in *South African Historical Journal*, 16:1, 1984, pp. 82–83, http://www.tandfonline.com/doi/pdf/10.1080/02582478408671588, accessed 5 December 2016.

Wegener, A., *The Origins of Continents and Oceans*, John Biram, (translator), New York, Dover Publications Inc, 1966.

Whitehouse, D., *Journey to the Centre of the Earth: A Scientific Exploration into the Heart of Our Planet*, London, Weidenfeld and Nicholson, 2016.

Wilson, D.B., *Seeking Nature's Logic: Natural Philosophy in the Scottish Enlightenment.* University Park, Pennsylvania State University Press, 2009.

Williams, W., "Introduction" in *Geology of the Appalachian-Caledonian Orogen in Canada and Greenland*, Harold Williams, (ed), Geological Society of America and Geological Survey of Canada, 1994.

Willoughby, M., "Overview of Hallet Cove Geological History" https://www.mindat.org/article.php/957/Overview+of+Hallett+Cove+Geological+History, accessed 18 October 2017.

Wilson, L.G., *Charles Lyell: The Years to 1841: The Revolution in Geology*, New York and London, Yale University Press, 1972.

Wood, R.M., *The Dark Side of the Earth*, London, George Allen and Unwin Publishers Ltd, 1985.

Worger, W.H., *South Africa's City of Diamonds: Mine Workers and Monopoly Capitalism in Kimberley, 1867–1895*, Craighall, AD Donker, 1987.
Lisa Yount, Y., *Alfred Wegener: Creator of the Continental Drift Theory*, New York, Chelsea House, 2009.
Zeller, S., "The Colonial World as Geological Metaphor: Strata[gems] of Empire in Victorian Canada" in *Osiris*, 2nd Series, Vol 15, "Nature and Empire: Science and the Colonial Enterprise", 2000, p. 105.

Index[1]

[1] Note: Page numbers followed by 'n' refer to notes.

S. Chetty, *'Africa Forms the Key'*, Cambridge Imperial and Post-Colonial Studies, https://doi.org/10.1007/978-3-030-52711-2

N

O

P

The manufacturer's authorised representative in the EU is Springer Nature Customer Service Centre GmbH, Europaplatz 3, 69115 Heidelberg, Germany. If you have any concerns regarding our products, please contact ProductSafety@springernature.com

Printed and bound by CPI Group (UK) Ltd, Croydon, CR0 4YY
30/09/2026
02225195-0003